WHOLE NUMBERS: MULTIPLICATION AND DIVISION

Connie Eichhorn
Mary Garland

Power Math Series

CAMBRIDGE Adult Education
A Division of Simon & Schuster
Upper Saddle River, New Jersey

Dr. Connie Eichhorn is the Supervisor of Transitional Programs for the Omaha Public Schools. She is the former president of the American Association of Adult and Continuing Education. Dr. Eichhorn is very active in adult education and has consulted in the development of a variety of adult education materials.

AUTHOR: Mary Garland

EXECUTIVE EDITOR: Mark Moscowitz

EDITOR: Kirsten Richert

PRODUCTION DIRECTOR: Penny Gibson

PRODUCTION EDITOR: Linda Greenberg

PRINT BUYER: Patricia Alvarez

ART DIRECTOR: Marianne Frasco

BOOK DESIGN: Patrice Sheraton

ELECTRONIC PAGE PRODUCTION: Curriculum Concepts

COVER DESIGN: Amy Rosen

COVER PHOTO: © David Bishop/Phototake

Printed in the United States of America

1 2 3 4 5 6 7 8 9 10 99 98 97 96 95

ISBN 0-13-078841-4

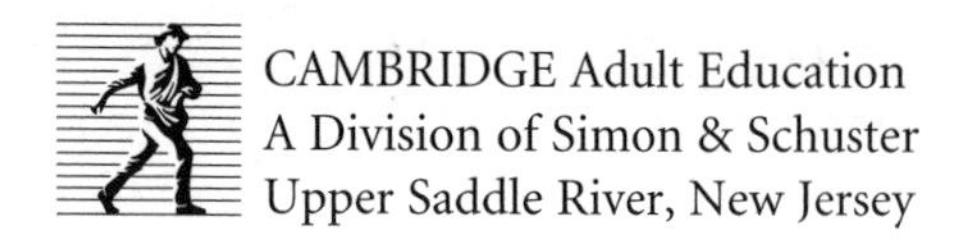

Contents

To the Learner

The ten books in the Power Math series are designed to help you understand and practice arithmetic skills. Lessons are easy to use and the problems are designed to address every-day adult life.

Lessons have the following features:

- Every lesson begins with a sample problem from real-life experience. You are asked to use your knowledge of math to find a solution.
- The *Think* section takes you through the thought process you might use to organize the information in the problem and choose a problem-solving approach.
- The *Do* section shows you step-by-step how to solve the problem.
- In *Try These*, you will warm up by solving a few problems similar to the opening sample problem. Some steps are worked for you to get you off to a good start.
- The *Practice* section gives you ample opportunities to practice the skill presented in the lesson.
- The *Solving Problems* section applies the math skill in a practical application from your experience as a consumer and worker.

Within each book, review lessons give you opportunities to decide whether you have mastered the skills presented in the book. The *Answer Key* section at the end of the book has answers and worked-out solutions for the problems in the book. Use the answers to check your work. Use the worked-out solutions to make sure your approach to a problem was the correct one.

By working carefully through the exercises in this book, you will find increased confidence in your math skills. Good luck.

WHOLE NUMBERS: MULTIPLICATION AND DIVISION

Lyn has three packs of soda. There are six cans in each pack. How many cans of soda does Lyn have all together?

Think

All together is a key phrase that tells you to add or multiply.

You can add to find how many cans of soda Lyn has all together.

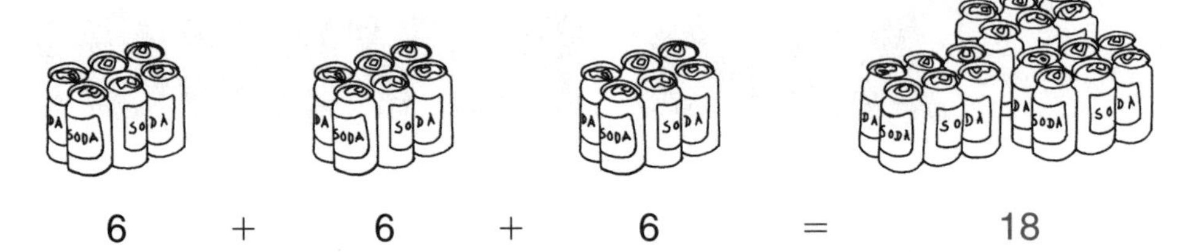

$6 + 6 + 6 = 18$

Or you can multiply. Multiplication is repeated addition. Multiplication is shown by a times symbol (×). The answer is called the product.

Do

To find how many cans of soda Lyn has, you need to add 6 three times. To find the answer using multiplication, multiply 6 times 3.

$6 \times 3 = 18$

Lyn has 18 cans of soda.

NOTE: Multiplication problems can be written in a row. $\mathbf{6 \times 3 = 18}$
Multiplication problems can also be written in a column.

$$\begin{array}{r} \mathbf{6} \\ \underline{\times\, \mathbf{3}} \\ \mathbf{18} \end{array}$$

The order in which the numbers are written does not change the answer.

$6 \times 3 = 18$
$3 \times 6 = 18$

Multiplying a number by 0 is 0. $\mathbf{4 \times 0 = 0}$
Multiplying a number by 1 is that number. $\mathbf{4 \times 1 = 4}$

Try These

Add. Then multiply.

1. $5 + 5 + 5 =$ _____
$5 \times 3 =$ _____

2. $2 + 2 + 2 + 2 =$ _____
$2 \times 4 =$ _____

PRACTICE

Add. Then multiply

3. $2 + 2 + 2 =$ _____
$2 \times 3 =$ _____

4. $4 + 4 + 4 + 4 =$ _____
$4 \times 4 =$ _____

5. $6 + 6 =$ _____
$6 \times 2 =$ _____

6. $9 + 9 + 9 + 9 =$ _____
$9 \times 4 =$ _____

Multiply.

7. $4 \times 9 =$ _____
$9 \times 4 =$ _____

8. $2 \times 8 =$ _____
$6 \times 2 =$ _____

9. $5 \times 7 =$ _____
$7 \times 5 =$ _____

10. $6 \times 9 =$ _____
$9 \times 6 =$ _____

11. $0 \times 2 =$ _____
$2 \times 0 =$ _____

12. $6 \times 3 =$ _____
$3 \times 6 =$ _____

13. $1 \times 3 =$ _____
$3 \times 1 =$ _____

14. $7 \times 8 =$ _____
$8 \times 7 =$ _____

Key Words for Multiplication

all together, by, entire, in all, of, product, total, twice, times, whole

Look for these key words. They are hints to help you decide to multiply. You may not find a key word in every problem.

Solving Problems

Solve. Look for key words that tell you to multiply.

15. Richard bought 5 trays of tomato plants. Each tray has 6 plants. How many plants did Richard buy?

_____

16. Tram earns $8 per hour. If he works 2 hours, how much will he earn?

_____

Check your answers on page 73.

LESSON

The multiplication facts in this lesson are used in all multiplication problems. It is important to know all the basic multiplication facts.

Think

You can use the Multiplication Table below to find the answers to all the basic multiplication facts.

Do

Find the product of 5×7.

×	0	1	2	3	4	5	6	7	8	9
0	0	0	0	0	0	0	0	0	0	0
1	0	1	2	3	4	5	6	7	8	9
2	0	2	4	6	8	10	12	14	16	18
3	0	3	6	9	12	15	18	21	24	27
4	0	4	8	12	16	20	24	28	32	36
5	0	5	10	15	20	25	30	35	40	45
6	0	6	12	18	24	30	36	42	48	54
7	0	7	14	21	28	35	42	49	56	63
8	0	8	16	24	32	40	48	56	64	72
9	0	9	18	27	36	45	54	63	72	81

Step 1. Locate 5 in the top row. Locate 7 in the left-hand column.

Step 2. Move down column 5 until you reach row 7. Move across row 7 until you reach column 5.

Step 3. The box where these numbers meet is 35.
The product of 5×7 is 35.

Try These

Use the Multiplication Table to find these answers.

1. $\begin{array}{r} 2 \\ \times 7 \\ \hline \end{array}$ $\begin{array}{r} 7 \\ \times 2 \\ \hline \end{array}$

2. $\begin{array}{r} 4 \\ \times 8 \\ \hline \end{array}$ $\begin{array}{r} 8 \\ \times 4 \\ \hline \end{array}$

3. $3 \times 5 =$ _____
$5 \times 3 =$ _____

4. $9 \times 6 =$ _____
$6 \times 9 =$ _____

PRACTICE

Multiply. Use Multiplication Table if necessary.

5.	4	7	0	3	5	1	9	8	6	2
	× 0	× 0	× 0	× 0	× 0	× 0	× 0	× 0	× 0	× 0
6.	5	2	6	4	0	8	7	3	9	1
	× 1	× 1	× 1	× 1	× 1	× 1	× 1	× 1	× 1	× 1
7.	2	8	4	7	5	6	1	0	9	3
	× 2	× 2	× 2	× 2	× 2	× 2	× 2	× 2	× 2	× 2
8.	4	6	5	2	9	7	1	3	8	0
	× 3	× 3	× 3	× 3	× 3	× 3	× 3	× 3	× 3	× 3
9.	0	9	3	7	4	8	2	6	1	5
	× 4	× 4	× 4	× 4	× 4	× 4	× 4	× 4	× 4	× 4
10.	7	4	6	0	1	9	8	5	3	2
	× 5	× 5	× 5	× 5	× 5	× 5	× 5	× 5	× 5	× 5
11.	9	3	8	6	1	0	5	7	4	2
	× 6	× 6	× 6	× 6	× 6	× 6	× 6	× 6	× 6	× 6
12.	3	5	2	0	6	8	7	9	1	4
	× 7	× 7	× 7	× 7	× 7	× 7	× 7	× 7	× 7	× 7
13.	5	1	6	8	4	0	3	9	7	2
	× 8	× 8	× 8	× 8	× 8	× 8	× 8	× 8	× 8	× 8
14.	2	7	0	6	5	8	1	3	4	9
	× 9	× 9	× 9	× 9	× 9	× 9	× 9	× 9	× 9	× 9

Solving Problems

Fill in the blanks with a basic multiplication fact. There is more than one pair of numbers that will make a correct answer.

15. _____ × _____ = 6   **16.** _____ × _____ = 45   **17.** _____ × _____ = 9

18. _____ × _____ = 5   **19.** _____ × _____ = 10   **20.** _____ × _____ = 24

Check your answers on pages 73-74.

The attendance at a Colorado Rockies game was 70,342. Which digit shows how many ten thousands of people attended the game?

Think

Each place value column has a name. To find the value of a digit, figure out which place value column the digit is in. A place value chart shows the name of each column.

Do

Write the number on the place value chart.
Find the digit in the ten thousands place.
In this number, the digit 7 represents 7 ten thousands or 70,000.

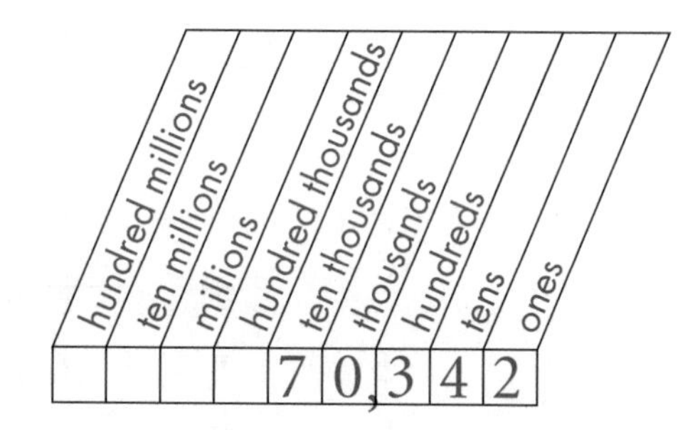

Try These

Write each number on the place value chart. Then write the place value of the circled digit.

1. **1.** ④,189 ______________
2. **2.** 912,⑧06 ______________
3. **3.** 2③0 ______________
4. **4.** 10,50⑧ ______________
5. **5.** ③,175 ______________
6. **6.** 1,⓪03,462 ______________

	hundred millions	ten millions	millions	hundred thousands	ten thousands	thousands	hundreds	tens	ones
1.						4,	1	8	9
2.									
3.									
4.									
5.									
6.									

PRACTICE

Build the numbers on the place value chart.

7. ones column 3
   tens column 6
   hundreds column 8
   thousands column 7
   ten thousands column 4

8. millions column 8
   ones column 3
   hundred thousands column 0
   thousands column 2
   tens column 9
   hundreds column 5
   ten thousands column 1

	hundred millions	ten millions	millions	hundred thousands	ten thousands	thousands	hundreds	tens	ones
7.						7			3
8.									

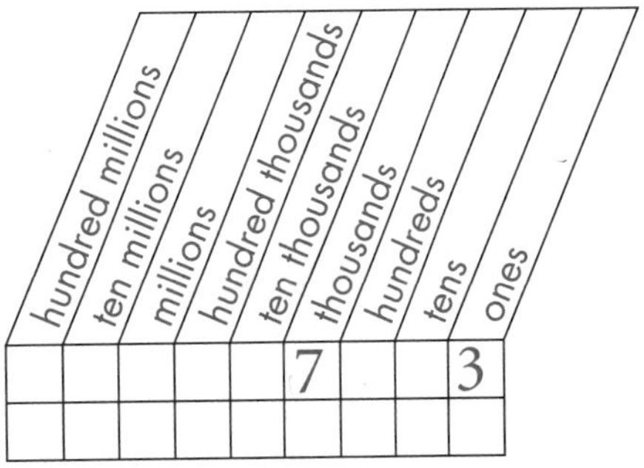

Read each number. Write the digit that is found in each place value column.

9. 159,362
   ones _____
   tens _____
   hundreds _____
   thousands _____
   ten thousands _____
   hundred thousands _____

10. 348,610
    hundreds _____
    ten thousands _____
    ones _____
    thousands _____
    tens _____
    hundred thousands _____

Solving Problems

Write the number of ones, tens, hundreds, and thousands in each number.

11. 7,385 = _____thousands _____hundreds _____tens _____ones

12. 4,911 = _____thousands _____hundreds _____tens _____ones

Match the numbers written in digits to the numbers in words.

13. 840,712 _____ a. eighty-four thousand, seven hundred twelve

14. 84,712 _____ b. eight hundred forty thousand, seven hundred twelve

15. 804,127 _____ c. eight hundred four thousand, one hundred twenty-seven

Check your answers on page 74.

Jewel rides the subway to and from work each day. The travel time between Burlington Station and Downtown Station is 34 minutes. How much time does she spend traveling the subway each day?

Think

To find how long Jewel travels on the subway each day, multiply the travel time between Burlington Station and Downtime Station by 2.

Do

Step 1. Line the numbers in a column.

Step 2. Multiply 2 by the digit in the ones column.

$2 \times 4 = 8$

Write 8 in the ones column.

$$\begin{array}{r} 34 \\ \times\ 2 \\ \hline 68 \end{array}$$

Step 3. Multiply 2 by the digit in the tens column.

$2 \times 3 = 6$

Write 6 in the tens column.

Jewel spends 68 minutes traveling the subway each day.

NOTE: If the top number of the multiplication problem is a three digit number, follow the same steps:

$$\begin{array}{r} 431 \\ \times\ 3 \\ \hline 1{,}293 \end{array}$$

Multiply the digit in the ones column. $3 \times 1 = 3$

Multiply the digit in the tens column. $3 \times 3 = 9$

Multiply the digit in the hundreds column. $3 \times 4 = 12$

Try These

Multiply. Line the numbers in a column if necessary.

1. $\begin{array}{r} 53 \\ \times 3 \\ \hline \end{array}$

2. $\begin{array}{r} 413 \\ \times 2 \\ \hline \end{array}$

3. $61 \times 5 =$ _____

4. $243 \times 2 =$ _____

PRACTICE

Multiply.

5. $\begin{array}{r} 32 \\ \times 3 \\ \hline \end{array}$

6. $\begin{array}{r} 15 \\ \times 1 \\ \hline \end{array}$

7. $\begin{array}{r} 83 \\ \times 2 \\ \hline \end{array}$

8. $\begin{array}{r} 51 \\ \times 5 \\ \hline \end{array}$

9. $\begin{array}{r} 61 \\ \times 3 \\ \hline \end{array}$

10. $\begin{array}{r} 41 \\ \times 2 \\ \hline \end{array}$

11. $\begin{array}{r} 12 \\ \times 3 \\ \hline \end{array}$

12. $\begin{array}{r} 32 \\ \times 4 \\ \hline \end{array}$

13. $\begin{array}{r} 124 \\ \times 2 \\ \hline \end{array}$

14. $\begin{array}{r} 621 \\ \times 4 \\ \hline \end{array}$

15. $\begin{array}{r} 211 \\ \times 6 \\ \hline \end{array}$

16. $\begin{array}{r} 334 \\ \times 2 \\ \hline \end{array}$

17. $42 \times 3 =$ _____

18. $31 \times 2 =$ _____

19. $22 \times 3 =$ _____

20. $22 \times 4 =$ _____

21. $211 \times 8 =$ _____

22. $312 \times 4 =$ _____

23. $54 \times 2 =$ _____

24. $134 \times 2 =$ _____

25. $421 \times 3 =$ _____

Solving Problems

Solve.

26. The cookie crisps Zeria takes in her lunch have 21 calories each. If she eats 5, how many calories is that in all?

_____

27. Irene has $50 to buy clothes for her sons. T-shirts are $4 each. Not counting sales tax, will the $50 be enough to buy 12 T-shirts?

_____

28. Dewey saves $122 from each monthly check. How much will he save in four months?

_____

Check your answers on page 75.

Carrying in One-Digit Multiplication

Maria is renting an apartment for $415 a month for the summer. What is her total rent for three months?

Think

Total is a key word that can tell you to multiply. To find the total rent for three months, multiply the monthly rent by 3.

Do

Step 1. Line the numbers in a column.

Step 2. Multiply 3 by the digit in the ones column.
$3 \times 5 = 15$ (Think 15 = 1 ten and 5 ones.)
Write 5 in the ones column.
Carry 1 to the tens column.

$$\begin{array}{r} {}^{1} \\ \$415 \\ \times\ 3 \\ \hline \$1{,}245 \end{array}$$

Step 3. Multiply 3 by the digit in the tens column.
$3 \times 1 = 3$
Add the "carried" 1 to 3. ($1 + 3 = 4$)
Write 4 in the tens column.

Step 4. Multiply 3 by the digit in the hundreds column.
$3 \times 4 = 12$
Write 2 in the hundreds column.
Write 1 in the thousands column.

Maria will pay $1,245 for three months' rent.

Try These

Multiply. Line the numbers in a column if necessary.

1. $\begin{array}{r} 24 \\ \times\ 5 \\ \hline \end{array}$ **2.** $\begin{array}{r} 613 \\ \times\ 4 \\ \hline \end{array}$ **3.** $72 \times 6 =$ _____ **4.** $413 \times 2 =$ _____

Multiply.

5. 18×8	6. 87×6	7. 129×5	8. 102×7
9. 255×7	10. 496×8	11. 538×4	12. 32×6
13. 96×5	14. 64×6	15. 747×9	16. 523×4

17. $34 \times 3 =$ _____ **18.** $48 \times 8 =$ _____ **19.** $214 \times 7 =$ _____

20. $672 \times 4 =$ _____ **21.** $139 \times 6 =$ _____ **22.** $763 \times 9 =$ _____

Solving Problems

Solve.

23. Doug is sending his parents $115 a month to pay back a loan. How much of the loan will be paid off in 9 months?

____

24. Geraldine has a beauty shop in her home. She schedules 25 appointments a week. How many appointments does she schedule in four weeks?

____

25. Jenjuro and Deanne deliver 53 morning papers every day of the week. How many papers do they deliver all together in one week?

____

Check your answers on page 75.

On Sunday the Bagel Bin sells bags filled with bagels. Each bag contains a "baker's dozen." This morning 32 customers each bought a bag. How many bagels were sold in all? (HINT: a baker's dozen = 13 bagels)

Think

Each customer bought 13 bagels. *In all* tells you to multiply. To find the number of bagels sold in all, multiply the number of customers who bought bagels by 13.

Do

Step 1. Line the numbers in a column.

Step 2. Multiply 3 by 32 to get the first partial product.
$3 \times 2 = 6$
Write 6 in the ones column.
$3 \times 3 = 9$
Write 9 in the tens column.

$$\begin{array}{r} 32 \\ \times\ 13 \\ \hline 96 \end{array}$$

Step 3. Put 0 in the ones column before multiplying for the next partial product.

Step 4. Multiply 1 by 32 to get the next partial product.
$1 \times 2 = 2$
Write 2 in the tens column.
$1 \times 3 = 3$
Write 3 in the tens column.

$$\begin{array}{r} 32 \\ \times\ 13 \\ \hline 96 \\ 320 \end{array}$$

Step 5. Add the partial products.
$96 + 320 = 416$

$$\begin{array}{r} 32 \\ \times\ 13 \\ \hline 96 \\ +\ 320 \\ \hline 416 \end{array}$$

The Bagel Bin sold 416 bagels in all.

Try These

Multiply. Line the numbers in a column if necessary.

1. $\begin{array}{r} 23 \\ \times\ 31 \\ \hline \end{array}$

2. $\begin{array}{r} 421 \\ \times\ 42 \\ \hline \end{array}$

3. $32 \times 13 =$ _____

4. $121 \times 23 =$ _____

PRACTICE

Multiply.

5. $\begin{array}{r} 32 \\ \times\ 31 \\ \hline \end{array}$

6. $\begin{array}{r} 413 \\ \times\ 32 \\ \hline \end{array}$

7. $\begin{array}{r} 24 \\ \times\ 12 \\ \hline \end{array}$

8. $\begin{array}{r} 111 \\ \times\ 36 \\ \hline \end{array}$

9. $\begin{array}{r} 124 \\ \times\ 22 \\ \hline \end{array}$

10. $\begin{array}{r} 432 \\ \times\ 13 \\ \hline \end{array}$

11. $\begin{array}{r} 332 \\ \times\ 23 \\ \hline \end{array}$

12. $\begin{array}{r} 42 \\ \times\ 21 \\ \hline \end{array}$

13. $131 \times 32 =$ _____

14. $222 \times 41 =$ _____

15. $13 \times 21 =$ _____

16. $423 \times 33 =$ _____

17. $34 \times 12 =$ _____

18. $141 \times 21 =$ _____

Solving Problems

Solve.

19. Each employee at the Mako Plant will get a raise of $24 a month. What will the yearly raise be for each employee at the Mako Plant?

_____

20. Josephine is making a quilt which calls for 33 blocks of fabric in each panel. There are 12 panels in the quilt. How many blocks of fabric will Josephine need to cut for the quilt?

_____

21. The Greenview Ecology Club planted 312 rows of seedlings after the forest fire. They planted 42 seedlings in each row. How many seedlings did they plant?

_____

Check your answers on pages 75-76.

The Superior Driving School promises that every student will get at least 240 minutes of driving time with an instructor. Forty-seven students are signed up for the summer session. During the summer, how many minutes will the driving instructors spend riding with student drivers?

Think

To find how many minutes the instructors will spend riding with student drivers, multiply 240 minutes by 47, the number of students in the summer session.

Do

Step 1. Line the numbers in a column.

Step 2. Multiply 7 by 240 to get the first partial product. Remember to "carry" to the next column if necessary.
$7 \times 240 = 1{,}680$

$$\begin{array}{r} 240 \\ \times\ 47 \\ \hline 1{,}680 \end{array}$$

Step 3. Place 0 in the ones column before multiplying for the next partial product.

Step 4. Multiply 4 by 240 to get the next partial product. Remember to "carry" to the next column if necessary.
$4 \times 240 = 960$
The next partial product is 9,600.

$$\begin{array}{r} 240 \\ \times\ 47 \\ \hline 1{,}680 \\ 9{,}600 \end{array}$$

Step 5. Add the partial products.
$1{,}680 + 9{,}600 = 11{,}280$

$$\begin{array}{r} 240 \\ \times\ 47 \\ \hline 1{,}680 \\ +\ 9{,}600 \\ \hline 11{,}280 \end{array}$$

The driving instructors will spend 11,280 minutes riding with student drivers.

Try These

Multiply. Line the numbers in a column if necessary.

1. 56×75
2. 149×32
3. $74 \times 19 =$ _____
4. $327 \times 72 =$ _____

PRACTICE

Multiply.

5. 39×54
6. 435×74
7. 24×33
8. 56×28
9. 18×16
10. 94×93
11. 852×28
12. 376×49
13. $47 \times 67 =$ _____
14. $344 \times 75 =$ _____
15. $98 \times 94 =$ _____
16. $596 \times 29 =$ _____
17. $25 \times 84 =$ _____
18. $72 \times 35 =$ _____

Solving Problems

Solve.

19. A mechanic charged $67 each to replace 16 tires on driver training cars. What was the total cost to replace the tires?

_____

20. Thirty-nine students took the written part of the driving test. The test had 54 questions. How many answers did the driving instructor check all together?

_____

21. Each training car averages 28 miles per gallon of gasoline. On average, how many miles could one of the cars drive on 18 gallons of gas?

_____

22. Tuition at Superior Driving School is $95 per student. How much did the Superior Driving School collect in tuition for 47 students?

_____

Check your answers on page 76.

LESSON 8

Carrying in Three-Digit Multiplication

Midwest Air has 125 flights per day. Each flight can hold 255 passengers. How many passengers can the airlines carry in a day?

Think

To find the number of passengers, multiply the number of flights by the number of passengers on one flight.

Do

Step 1. Line the numbers in a column.

Step 2. Multiply 5 by 255 to get the first partial product. Remember to "carry" to the next column if necessary.
$5 \times 255 = 1{,}275$

$$\begin{array}{r} 255 \\ \times\ 125 \\ \hline 1{,}275 \end{array}$$

Step 3. Place 0 in the ones column before multiplying for the next partial product.

Step 4. Multiply 2 by 255 to get the next partial product. Remember to "carry" to the next column if necessary.
$2 \times 255 = 510$
The next partial product is 5,100.

$$\begin{array}{r} 255 \\ \times\ 125 \\ \hline 1{,}275 \\ 5{,}100 \end{array}$$

Step 5. Put 0 in the ones and tens column before multiplying for the next partial product.

Step 6. Multiply 1 by 255 to get the next partial product.
Remember to "carry" to the next column if necessary.
$1 \times 255 = 255$
The next partial product is 25,500.

$$\begin{array}{r} 255 \\ \times\ 125 \\ \hline 1{,}275 \\ 5{,}100 \\ 25{,}500 \\ \hline 31{,}875 \end{array}$$

Step 7. Add the partial products.
$1{,}275 + 5{,}100 + 25{,}500 = 31{,}875$

Midwest Air can carry 31,875 passengers per day.

Try These

Multiply. Line the numbers in a column if necessary.

1. $\begin{array}{r} 271 \\ \times\ 842 \\ \hline \end{array}$ 2. $\begin{array}{r} 526 \\ \times\ 715 \\ \hline \end{array}$ 3. $382 \times 145 =$ _____ 4. $367 \times 418 =$ _____

PRACTICE

Multiply.

5. $\begin{array}{r} 452 \\ \times\ 133 \\ \hline \end{array}$ 6. $\begin{array}{r} 421 \\ \times\ 638 \\ \hline \end{array}$ 7. $\begin{array}{r} 213 \\ \times\ 592 \\ \hline \end{array}$

8. $\begin{array}{r} 564 \\ \times\ 581 \\ \hline \end{array}$ 9. $\begin{array}{r} 564 \\ \times\ 724 \\ \hline \end{array}$ 10. $\begin{array}{r} 912 \\ \times\ 238 \\ \hline \end{array}$

11. $988 \times 112 =$ _____ 12. $417 \times 873 =$ _____ 13. $292 \times 634 =$ _____

14. $949 \times 217 =$ _____ 15. $367 \times 418 =$ _____ 16. $428 \times 782 =$ _____

Solving Problems

Solve.

17. Buyers for a children's shoe store ordered 325 pairs of sneakers for each of its 114 stores. How many pairs of shoes did the buyers order?

_____

18. Shanti waits tables at a diner. She figures that the diner averages 115 customers during her shift. If Shanti works 255 days per year, how many customers will she see?

_____

19. Pizza-to-Go delivers 175 pizzas per day. How many pizzas will they deliver in 181 days, the first six months of the year?

_____

Check your answers on page 76.

LESSON 9 Multiplying With Zero

The Community Center is taking pledges. So far, there are 206 pledges of $25. How much will the center collect?

Think

To find the amount of pledge money, multiply the number of pledges by $25. Remember, multiplying any number by 0 is 0.

Do

Step 1. Line the numbers in a column.

Step 2. Multiply 5 by 206 to get the first partial product.
Multiply 5 by the digit in the ones column. ($5 \times 6 = 30$)
"Carry" 3 to the tens column.
Multiply 5 by the digit in the tens column. ($5 \times 0 = 0$)
Add the "carried" 3. ($3 + 0 = 3$)
Multiply 5 by the digit in the hundreds column.
($5 \times 2 = 10$)
$206 \times 5 = 1{,}030$
The partial product is 1,030.

$$\begin{array}{r} {\scriptstyle 3} \\ 206 \\ \times\ \$25 \\ \hline 1{,}030 \end{array}$$

Step 3. Place 0 in the ones column before multiplying for the next partial product.

Step 4. Multiply 2 by 206 to get the next partial product.
Multiply 2 by the digit in the ones column. ($2 \times 6 = 12$)
"Carry" 1 to the tens column.
Multiply 2 by the digit in the tens column. ($2 \times 0 = 0$)
Add the "carried" 1. ($1 + 0 = 1$)
Multiply 2 by the digit in the hundreds column.
($2 \times 2 = 4$)
$206 \times 2 = 412$
The next partial product is 4,120.

$$\begin{array}{r} {\scriptstyle 1} \\ 206 \\ \times\ \$25 \\ \hline 1{,}030 \\ 4{,}120 \end{array}$$

Step 5. Add the partial products.
$1{,}030 + 4{,}120 = 5{,}150$

$$\begin{array}{r} 206 \\ \times\ \$25 \\ \hline 1{,}030 \\ +\ 4{,}120 \\ \hline \$5{,}150 \end{array}$$

The center will collect $5,150 in pledges.

Try These

Multiply. Line the numbers in a column if necessary.

1. 607×52

2. $8{,}003 \times 28$

3. $406 \times 36 =$ _____

4. $2{,}038 \times 19 =$ _____

PRACTICE

Multiply.

5. 705×65

6. $6{,}020 \times 51$

7. $3{,}007 \times 41$

8. 805×62

9. $1{,}011 \times 84$

10. $1{,}806 \times 46$

11. $6{,}005 \times 23$

12. $4{,}023 \times 35$

13. $310 \times 24 =$ _____

14. $706 \times 29 =$ _____

15. $2{,}050 \times 33 =$ _____

16. $6{,}005 \times 45 =$ _____

17. $3{,}082 \times 16 =$ _____

18. $940 \times 76 =$ _____

Solving Problems

Solve.

19. Monte plans to paint a house that should take about 36 hours to paint. Would Monte see more profit if he was paid $12 an hour for 36 hours or if he was paid a flat fee of $380?

_____

20. One hundred ten businesses each donated 3 dozen items to the food pantry. How many food items were donated to the food pantry by all the businesses?
(HINT: 1 dozen = 12)

_____

21. Big Screen Video can stock 150 videos in one display rack. How many videos can be stocked in 25 display racks?

_____

Check your answers on pages 76-77.

Tamika won a lottery. She will get $1,000 each month for the next 36 months. How much will she get in all?

Think

To find how much Tamika will get in all, multiply 36 by $1,000. You can use a shortcut to multiply by 1,000.

Do

Step 1. Line the numbers in a column.

Step 2. Write 3 zeros in the product.

Step 3. Multiply 1 by 36.
$1 \times 36 = 36$

$$\begin{array}{r} 36 \\ \times\ \$1000 \\ \hline \$36{,}000 \end{array}$$

Tamika will get $36,000 in 36 months.

Use this shortcut when multiplying by 10, 100, 1000.

To multiply a number by 10, write 1 zero in the product and then multiply the other digits by 1.

To multiply a number by 100, write 2 zeros in the product and then multiply the other digits by 1.

To multiply a number by a 1,000, write 3 zeros in the product and then multiply the other digits by 1.

$$\begin{array}{r} 338 \\ \times\ 10 \\ \hline 3{,}380 \end{array} \qquad \begin{array}{r} 685 \\ \times\ 100 \\ \hline 68{,}500 \end{array} \qquad \begin{array}{r} 421 \\ \times\ 1{,}000 \\ \hline 421{,}000 \end{array}$$

Did you notice? Each time you wrote the same number of zeros as there were in the number you multiplied by.

10 → 1 zero
100 → 2 zeros
1,000 → 3 zeros

Try These

Multiply. Line the numbers in a column if necessary.

1. 49×100
2. 671×10
3. $152 \times 1{,}000 =$ _____
4. $830 \times 10 =$ _____

PRACTICE

Multiply.

5. 92×10
6. 543×10
7. 21×10
8. 436×10
9. 184×100
10. 3×100
11. $1{,}562 \times 100$
12. 54×100
13. $873 \times 1{,}000$
14. $773 \times 1{,}000$
15. $60 \times 1{,}000$
16. $892 \times 1{,}000$
17. $596 \times 100 =$ _____
18. $3 \times 10 =$ _____
19. $96 \times 1{,}000 =$ _____
20. $644 \times 1{,}000 =$ _____
21. $875 \times 10 =$ _____
22. $21 \times 100 =$ _____

Solving Problems

Solve.

23. On an assembly line, James puts 1,000 paper clips in each box. How many paper clips does he need to fill 250 boxes?

_____

24. Clyde gave each of his grandchildren $10 for their birthdays last year. How much did he spend on his 13 grandchildren?

_____

25. The Market Co-op sells 100 bushels of fruits and vegetables each weekend. In 8 weekends, how many bushels are sold?

_____

Check your answers on page 77.

LESSON

11 Multiplying By Numbers Ending in Zero

Eight people from each of the 50 states attended a meeting in Memphis. How many people were at the meeting?

Think

To find how many people were at the meeting multiply 8 by 50. You can use a shortcut when multiplying by a number ending in zero.

Do

Step 1. Line the numbers in a column.

Step 2. Write 1 zero in the product.

Step 3. Multiply 5 by 8.
$5 \times 8 = 40$

$$\begin{array}{r} 8 \\ \times\ 50 \\ \hline 400 \end{array}$$

There were 400 people at the meeting.

Use this shortcut when multiplying by any number ending in zero.
Count the number of zeros at the end of the number. Write that same number of zeros in the product.
Multiply the remaining digits.
For example:

$$\begin{array}{r} 827 \\ \times\ 300 \\ \hline 248{,}100 \end{array} \qquad \begin{array}{r} 164 \\ \times\ 210 \\ \hline 1{,}640 \\ +\ 32{,}800 \\ \hline 34{,}440 \end{array}$$

Write 1 zero in the partial product and multiply by 1.

For the next partial product, write 0 in the ones and tens columns; then multiply by 2. Add the partial products.

Try These

Multiply. Line the numbers in a column if necessary.

1. $\begin{array}{r} 321 \\ \times\ 180 \\ \hline \end{array}$ **2.** $\begin{array}{r} 42 \\ \times\ 20 \\ \hline \end{array}$ **3.** $162 \times 110 =$ _____ **4.** $38 \times 500 =$ _____

Multiply.

5. 53 × 90

6. 126 × 40

7. 61 × 700

8. 307 × 50

9. 44 × 120

10. 75 × 80

11. 28 × 130

12. 98 × 60

13. 781 × 400 = _____

14. 87 × 50 = _____

15. 9 × 40 = _____

16. 33 × 20 = _____

17. 104 × 550 = _____

18. 868 × 900 = _____

Solving Problems

Solve.

19. There are 60 minutes in an hour and 24 hours in a day. How many minutes are there in a day?

_____

20. Shirley drives her ice cream truck 50 miles a day. She needs to service her truck after 1,500 more miles. Will she be able to drive her truck 3 days, 30 days, or 300 days before she brings it in for service?

_____

21. The capacity of the River Boat Casino is 850 people. The boat cruises the river 225 days a year and is always filled to capacity. Fill in the display on the billboard to show the number of people who ride the river boat each year.

_____

Check your answers on pages 77-78.

12 Rounding and Estimating in Multiplication

Ivory typed an essay on the computer. The essay has 18 lines with about 12 words per line. About how many words are in her essay?

Think

About is a key word that tells you to estimate. To find about how many words are in Ivory's essay, round 18 and 12 to their leading digits and then multiply.

Do

Step 1. Line the numbers in a column.

Step 2. Round each number to the leading digit.
If the digit to the right of the leading digit is 0, 1, 2, 3, or 4, leave the leading digit as it is.
If the digit to the right of the leading digit is 5, 6, 7, 8, or 9, increase the leading digit by one.
Round 18 → 20
Round 12 → 10

round

$$\begin{array}{r@{\;\rightarrow\;}r} 18 & 20 \\ \times\ 12 & \underline{10} \\ & 200 \end{array}$$

Step 3. Multiply the rounded numbers.
$20 \times 10 = 200$
Ivory's essay is about 200 words.

Try These

Round and estimate.

round

1. $\begin{array}{r} 27 \\ \times\ 63 \\ \hline \end{array}$

2. $\begin{array}{r} 593 \\ \times\ 141 \\ \hline \end{array}$

Round and estimate. Then multiply for an exact answer. Compare the exact answer and the estimate. Is the answer reasonable?

3. $\begin{array}{r} {}^{1}82 \\ \times\ 57 \\ \hline 574 \\ 4{,}100 \end{array}$ $\begin{array}{l} \rightarrow 80 \\ \rightarrow \underline{60} \end{array}$

4. $\begin{array}{r} 2{,}483 \\ \times\ 367 \\ \hline \end{array}$

PRACTICE

Round and estimate.

5. $\begin{array}{r} 409 \\ \times\ 51 \\ \hline \end{array}$

6. $\begin{array}{r} 829 \\ \times\ 72 \\ \hline \end{array}$

7. $\begin{array}{r} 38 \\ \times\ 67 \\ \hline \end{array}$

8. $\begin{array}{r} 374 \\ \times\ 95 \\ \hline \end{array}$

9. $\begin{array}{r} 33 \\ \times\ 25 \\ \hline \end{array}$

10. $\begin{array}{r} 724 \\ \times\ 41 \\ \hline \end{array}$

11. $\begin{array}{r} 5{,}156 \\ \times\ 613 \\ \hline \end{array}$

12. $\begin{array}{r} 979 \\ \times\ 253 \\ \hline \end{array}$

Round and estimate. Then multiply for an exact answer. Make sure your answer is reasonable.

13. 96 × 32 = _____

14. 21 × 47 = _____

15 109 × 26 = _____

16. 862 × 51 = _____

Solving Problems

Solve.

17. A leaking faucet drips about 13 ounces of water in an hour. About how many ounces are wasted in a day?
(HINT: There are 24 hours in one day.)

____

18. Sandy and her husband plan to put flyers on the cars in the parking lot to announce their new housecleaning business. There are 18 parking rows in the lot with 52 parking spots in each row. About how many flyers will they need?

____

Check your answers on page 78.

LESSON

Multiplication Review

1. $6 \times 4 =$ _____
2. $9 \times 3 =$ _____
3. $2 \times 1 =$ _____
4. $5 \times 0 =$ _____

5. 47×8
6. 26×3
7. 13×23
8. 42×24

9. 38×47
10. 64×30
11. 736×22
12. 286×67

13. 222×30
14. 348×500
15. 900×639
16. $2{,}006 \times 428$

17. 48×16
18. 387×21
19. 402×383
20. 261×94

21. 300×20
22. 492×770
23. 12×400
24. 75×810

Round and estimate. Then multiply for the exact answer. Make sure your answer is reasonable.

round

25. $\begin{array}{r} 62 \\ \times\ 47 \\ \hline \end{array}$

26. $\begin{array}{r} 92 \\ \times\ 38 \\ \hline \end{array}$

27. $\begin{array}{r} 56 \\ \times\ 19 \\ \hline \end{array}$

28. $\begin{array}{r} 107 \\ \times\ 83 \\ \hline \end{array}$

29. $\begin{array}{r} 663 \\ \times\ 521 \\ \hline \end{array}$

30. $\begin{array}{r} 491 \\ \times\ 76 \\ \hline \end{array}$

31. $\begin{array}{r} 64 \\ \times\ 15 \\ \hline \end{array}$

32. $\begin{array}{r} 514 \\ \times\ 439 \\ \hline \end{array}$

33. $\begin{array}{r} 287 \\ \times\ 64 \\ \hline \end{array}$

Solving Problems

Solve.

34. Kent is baking bread. The recipe call for 3 cups of wheat flour. He wants to double the recipe. How much wheat flour does he need?

____

35. A newborn baby sleeps about 16 to 18 hours a day. Approximately how much time does a baby sleep in the first month? (HINT: There are 30 days in a month.)

____

36. Emilio's van averages 28 miles to a gallon of gasoline. The van's gasoline tank holds 20 gallons of gasoline. Will Emilio be able to make a 400-mile trip on one tank of gasoline?

____

40. Eric gets paid $3 to finish detailing a car after it goes through the car wash. If Eric details 43 cars on Sunday, how much does he earn?

____

Solving Problems

Solve.

38. The Wu family pays $237 a month in rent. What is the amount of rent they pay in a year?

____

39. Nick swims 1,500 meters each day. How many meters does Nick swim during a 30-day month?

____

40. The Hispanic Heritage Museum does not charge admission, but collects about $61 in donations every day. How much does the museum, which is open daily, collect in one year? (HINT: There are 365 days in one year.)

____

Check your answers on pages 78-79.

Solving Problems With Addition, Subtraction, and Multiplication

Corey delivered balloon bouquets on Mother's Day. He was paid $5 an hour and worked for 12 hours. He also earned $37 in tips. What were his total earnings for that day?

Think

Using a problem-solving chart can help solve problems. The steps on the chart are:
Find the specific question the problem is asking.
Locate all the information you need.
Choose a method. Look for key words.
Work the problem using the information and the method.
Answer the question. (THINK: Is the answer reasonable?)

Do

Complete each step. Fill out the chart.
Question: Find the specific *question* the problem is asking.

Question	Information	Method	Process	Answer
What were his total earnings?				

Information: Locate all the *information* you need to solve the problem.

Question	Information	Method	Process	Answer
What were his total earnings?	paid $5 an hour worked 12 hours earned $37 tips			

Method: Look for key words or hints that tell you what operation to use to solve the problem. Sometimes the *method* may involve more than one operation.

Question	Information	Method	Process	Answer
What were his total earnings?	paid $5 an hour worked 12 hours earned $37 tips	Multiply Add		

Process: Work out the problem using your *information* and the *method.*

Question	Information	Method	Process	Answer
What were his total earnings?	paid $5 an hour worked 12 hours earned $37 tips	Multiply Add	12 × $5 = $60 $60 + 37 = $97	

Answer: Answer the question. (THINK: Is the answer reasonable?)

Question	Information	Method	Process	Answer
What were his total earnings?	paid $5 an hour worked 12 hours earned $37 tips	Multiply Add	12 × $5 = $60 $60 + 37 = $97	His total earnings were $97.

Yes. The answer is reasonable.

Try These

Solve using the chart below. Some require more than one step.

1. The Baskins bought a refrigerator for $719. They got $75 on a trade-in for their old refrigerator. How much was left for them to pay?

Question	Information	Method	Process	Answer
How much was left to pay?	cost of new refrigerator: $719 trade in: $75			

2. Magale took pictures at the Little League World Series. She took 3 rolls of 24 pictures and 2 rolls of 36 pictures. How many pictures did she take all together?

Question	Information	Method	Process	Answer

PRACTICE

Use the problem-solving chart to solve these problems.

3. Cohn Optical has a $15 rebate on eyeglasses. Akim wants a pair of glasses marked $89. What will be the cost of his glasses after the rebate?

Question	Information	Method	Process	Answer

4. The City Market usually has 20 check-out lines in service. Each cashier checks out about 18 customers in an hour. Estimate how many customers go through the check-out lines in one hour.

Question	Information	Method	Process	Answer

5. City Market has an express line on weekends. About 31 customers go through the express line every hour. How many customers go through the express line in 7 hours?

Question	Information	Method	Process	Answer

Solving Problems

Solve these problems using the steps from the problem-solving chart. You may draw the chart if you like.

6. Nita took her family to the zoo. She bought 2 adult tickets at $6 each and 3 children's tickets at $4 each. How much did she pay for the tickets?

____

7. Luis has two job offers. The first job pays $6 per hour for 40 hours of work per week. The second job pays $7 per hour for 35 hours of work per week. Which job will pay him more money?

____

8. Cecile has $45 to spend on clothes. She can buy denim pants for $16 a pair. How much will she have left if she buys two pairs?

____

Check your answers on pages 79-80.

Maggie is playing cards. There are 24 cards to be shared evenly among 4 players. How many cards will Maggie deal to each player?

Think

Shared is a key word that tells you to divide. To find how many cards Maggie will deal to each player, divide the number of cards by the number of players. Division is shown by a division symbol (÷) or a division bracket ($\overline{)}$). The answer is called the quotient.

Do

Division is repeated subtraction. When you divide, 4 into 24 you are finding out how many times you can subtract 4 from 24.

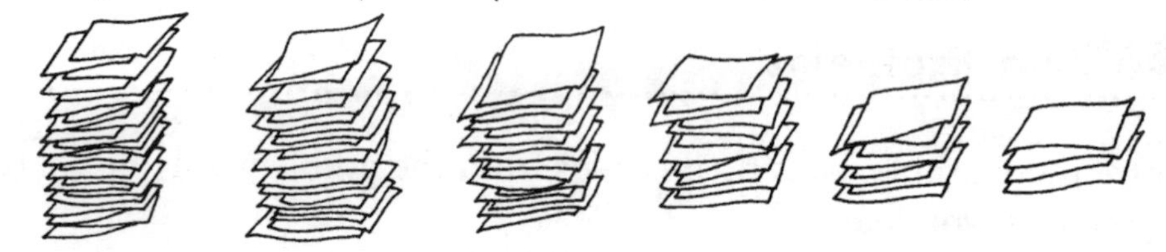

(24 − 4 = 20) (20 − 4 = 16) (16 − 4 = 12) (12 − 4 = 8) (8 − 4 = 4) (4 − 4 = 0)

To divide think how many times will 4 go into 24.

$$24 \div 4 = 6 \quad \text{or} \quad 4\overset{\displaystyle 6}{\overline{)24}}$$

Maggie will deal each player 6 cards.

NOTE: Changing the order of the numbers in the division problem changes the answer.

24 ÷ 4 does not equal 4 ÷ 24.

Dividing a number by 1 equals that number.

5 ÷ 1 = 5   20 ÷ 1 = 20

Dividing a number by 0 has no meaning. 5 cannot be divided by 0. Zero divided by any number is still 0.

Try These

Subtract. Then divide.

1. (9 − 3 = _____)   (6 − 3 = _____)   (3 − 3 = _____)
9 ÷ 3 = $3\overline{)9}$

2. (10 − 5 = _____)   (5 − 5 = _____)   10 ÷ 5 = $5\overline{)10}$

Subtract. Then divide.

3. $(8 - 2 = \_\_\_\_)$ $(6 - 2 = \_\_\_\_)$ $(4 - 2 = \_\_\_\_)$

$(2 - 2 = \_\_\_\_)$ $8 \div 2 =$ $2\overline{)8}$

4. $(15 - 5 = \_\_\_\_)$ $(10 - 5 = \_\_\_\_)$ $(5 - 5 = \_\_\_\_)$

$15 \div 5 =$ $5\overline{)15}$

5. $(24 - 8 = \_\_\_\_)$ $(16 - 8 = \_\_\_\_)$ $(8 - 8 = \_\_\_\_)$

$24 \div 8 =$ $8\overline{)24}$

Divide.

6. $35 \div 7 = \_\_\_\_$ **7.** $14 \div 2 = \_\_\_\_$ **8.** $27 \div 3 = \_\_\_\_$ **9.** $9 \div 9 = \_\_\_\_$

10. $21 \div 7 = \_\_\_\_$ **11.** $36 \div 6 = \_\_\_\_$ **12.** $5\overline{)25}$ **13.** $9\overline{)36}$

14. $4\overline{)28}$ **15.** $6\overline{)12}$ **16.** $4\overline{)24}$ **17.** $9\overline{)9}$

Solving Problems

Solve. Look for key words that tell you to divide.

18. How many groups of 7 are there in 42?

____

19. Eight players showed up for a game of basketball. Divide the players equally into two teams. How many players will be on each team?

____

Key Words for Multiplication

average, divide, equal parts, equally, groups, sets, share, split

Look for these key words. They are hints to help you decide to divide. You may not find a key word in every problem.

Check your answers on page 80.

LESSON

Division Facts

The division facts below are the basis for all division problems. Practice these division facts until you can solve them quickly and accurately.

Try These

Divide.

1. $1\overline{)7}$ $1\overline{)3}$ $1\overline{)8}$ $1\overline{)9}$ $1\overline{)1}$ $1\overline{)4}$ $1\overline{)5}$ $1\overline{)6}$ $1\overline{)0}$ $1\overline{)2}$

2. $2\overline{)8}$ $2\overline{)4}$ $2\overline{)14}$ $2\overline{)6}$ $2\overline{)10}$ $2\overline{)16}$ $2\overline{)0}$ $2\overline{)12}$ $2\overline{)18}$ $2\overline{)2}$

3. $3\overline{)9}$ $3\overline{)6}$ $3\overline{)15}$ $3\overline{)24}$ $3\overline{)12}$ $3\overline{)27}$ $3\overline{)21}$ $3\overline{)3}$ $3\overline{)18}$ $3\overline{)12}$

4. $4\overline{)36}$ $4\overline{)0}$ $4\overline{)20}$ $4\overline{)28}$ $4\overline{)12}$ $4\overline{)32}$ $4\overline{)16}$ $4\overline{)8}$ $4\overline{)24}$ $4\overline{)4}$

5. $5\overline{)5}$ $5\overline{)20}$ $5\overline{)35}$ $5\overline{)0}$ $5\overline{)45}$ $5\overline{)15}$ $5\overline{)40}$ $5\overline{)30}$ $5\overline{)25}$ $5\overline{)10}$

6. $6\overline{)18}$ $6\overline{)36}$ $6\overline{)12}$ $6\overline{)42}$ $6\overline{)6}$ $6\overline{)24}$ $6\overline{)48}$ $6\overline{)54}$ $6\overline{)0}$ $6\overline{)30}$

7. $7\overline{)0}$ $7\overline{)35}$ $7\overline{)28}$ $7\overline{)7}$ $7\overline{)42}$ $7\overline{)21}$ $7\overline{)49}$ $7\overline{)14}$ $7\overline{)56}$ $7\overline{)63}$

8. $8\overline{)72}$ $8\overline{)16}$ $8\overline{)0}$ $8\overline{)32}$ $8\overline{)8}$ $8\overline{)48}$ $8\overline{)56}$ $8\overline{)64}$ $8\overline{)40}$ $8\overline{)24}$

9. $9\overline{)27}$ $9\overline{)36}$ $9\overline{)54}$ $9\overline{)18}$ $9\overline{)81}$ $9\overline{)0}$ $9\overline{)72}$ $9\overline{)9}$ $9\overline{)63}$ $9\overline{)45}$

PRACTICE

Fill in the blanks.

10. $3 \times 2 =$ _____
$2 \times 3 =$ _____
$6 \div 3 =$ _____
$6 \div 2 =$ _____

11. $7 \times 8 =$ _____
$8 \times 7 =$ _____
$56 \div 7 =$ _____
$56 \div 8 =$ _____

12. $4 \times 9 =$ _____
$9 \times 4 =$ _____
$36 \div 4 =$ _____
$36 \div 9 =$ _____

13. $5 \times 1 =$ _____
$1 \times 5 =$ _____
$5 \div 5 =$ _____
$5 \div 1 =$ _____

14. $6 \times 4 =$ _____
$4 \times 6 =$ _____
$24 \div 6 =$ _____
$24 \div 4 =$ _____

15. $7 \times 5 =$ _____
$5 \times 7 =$ _____
$35 \div 7 =$ _____
$35 \div 5 =$ _____

16. $8 \times 3 =$ _____
$3 \times 8 =$ _____
$24 \div 8 =$ _____
$24 \div 3 =$ _____

17. $2 \times 6 =$ _____
$6 \times 2 =$ _____
$12 \div 2 =$ _____
$12 \div 6 =$ _____

18. $9 \times 6 =$ _____
$6 \times 9 =$ _____
$54 \div 9 =$ _____
$54 \div 6 =$ _____

19. $32 \div 8 =$ _____
$32 \div 4 =$ _____

20. $45 \div 5 =$ _____
$45 \div 9 =$ _____

21. $6 \div 2 =$ _____
$6 \div 3 =$ _____

22. $18 \div 6 =$ _____
$18 \div 3 =$ _____

23. $28 \div 7 =$ _____
$28 \div 4 =$ _____

24. $56 \div 8 =$ _____
$56 \div 7 =$ _____

25. $36 \div 9 =$ _____
$36 \div 4 =$ _____

26. $30 \div 6 =$ _____
$30 \div 5 =$ _____

27. $14 \div 7 =$ _____
$14 \div 2 =$ _____

Solving Problems

Fill in the blank with the number that makes the math fact correct.

28. $32 \div$ _____ $= 8$

29. $56 \div$ _____ $= 7$

30. $40 \div$ _____ $= 8$

31. $81 \div$ _____ $= 9$

32. $6 \div$ _____ $= 3$

33. $54 \div$ _____ $= 6$

34. $24 \div$ _____ $= 8$

35. $42 \div$ _____ $= 6$

36. $72 \div$ _____ $= 8$

37. $64 \div$ _____ $= 8$

38. $18 \div$ _____ $= 3$

39. $9 \div$ _____ $= 1$

40. $24 \div$ _____ $= 3$

41. $27 \div$ _____ $= 9$

42. $25 \div$ _____ $= 5$

Check your answers on pages 80-81.

Frank is packaging blank cassette tapes into boxes. He has 15 tapes left. Each box holds 5 tapes.
Write math facts to show:

The number of boxes needed to package 15 tapes.
The number of tapes that will be needed to fill 3 boxes.

Think

Knowing the multiplication facts can help you divide. Multiplication and division are opposite operations, so you can check a division answer by multiplying.

Do

Step 1. Boxes needed for 15 tapes.

$15 \div 5 = 3$ boxes needed

Step 2. Tapes needed to fill 3 boxes.

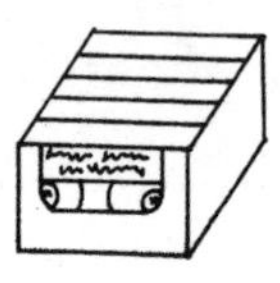
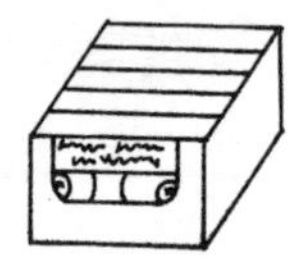

$3 \times 5 = 15$ tapes

Division is the opposite of multiplication. Check division by multiplication.

Division	**Multiplication**
$15 \div 5 = 3$ (5: divisor; 3: quotient)	$5 \times 3 = 15$

Multiply the divisor by the quotient. The answer should be the number you started with.

Try These

Use multiplication facts to help you divide.

1. $7 \times 6 = 42$
$42 \div 6 =$

2. $6 \times 5 =$ _____

Divide. Check by multiplication.

3. $16 \div 8 =$ _____ *check* $2 \times 8 =$ _____

4. $21 \div 3 =$ _____ *check*

PRACTICE

Use multiplication facts to help you divide.

5. $7 \times 7 = 49$
$49 \div 7 =$ _____

6. $9 \times 4 = 36$
$36 \div 4 =$ _____

7. $4 \times 3 =$ _____
$12 \div 3 =$ _____

8. $6 \times 5 = 30$
$30 \div 5 =$ _____

9. $2 \times 5 = 10$
$10 \div 5 =$ _____

10. $5 \times 8 =$
$40 \div 8 =$ _____

Divide. Check by multiplication.

11. $25 \div 5 =$ _____ *check*

12. $10 \div 2 =$ _____ *check*

13. $64 \div 8 =$ _____ *check*

14. $8 \div 1 =$ _____ *check*

15. $28 \div 4 =$ _____ *check*

16. $30 \div 6 =$ _____ *check*

17. $18 \div 3 =$ _____ *check*

18. $54 \div 9 =$ _____ *check*

19. $21 \div 7 =$ _____ *check*

Solving Problems

Solve.

20. Bob paid $30 for 6 tickets to a Dodger game.

a. Write a division fact to show how much each ticket cost.

____

b. Write a multiplication fact to show the cost of buying six $5 tickets.

____

Check your answers on page 82.

LESSON

18 One Digit Division

Kendall does volunteer work at the library. He needs to work 45 more hours to earn the Volunteer Award. He wants to divide the work evenly over the next 3 weeks. How many hours will he need to work each week?

Think

Divide evenly is a key phrase that tells you to divide. To find the number of hours Kendall needs to work each week, divide the total number of hours by the number of weeks.

Do

Step 1. Place the numbers in a division bracket.

$3\overline{)45}$

Step 2. Determine where the quotient will begin.

Since 4 can be divided by 3, the quotient will begin above the 4.

For every digit in the bracket, there is a digit above it in the quotient.

The answer will have two digits beginning above the 4.

$$\begin{array}{r} \_\,\_ \\ 3\overline{)45} \end{array}$$

Step 3. Use these steps to divide each number in the bracket by 3.

Divide. $4 \div 3 = 1$

Multiply. $1 \times 3 = 3$

Subtract. $4 - 3 = 1$

Bring down the next digit, 5.

$$\begin{array}{r} 1 \\ 3\overline{)45} \\ -\underline{3} \\ 15 \end{array}$$

Step 4. Repeat the process.

Divide. $15 \div 3 = 5$

Multiply. $5 \times 3 = 3$

Subtract. $15 - 15 = 0$

There are no other digits to bring down.

$$\begin{array}{r} 15 \\ 3\overline{)45} \\ -\underline{3} \\ 15 \\ -\underline{15} \\ 0 \end{array}$$

Kendall needs to work 15 hours each week for the next 3 weeks.

Try These

Divide. Place the numbers in a bracket if necessary.

1. $4\overline{)96}$ **2.** $5\overline{)865}$ **3.** 63 ÷ 3 =_____ **4.** 945 ÷ 7 =_____

PRACTICE

Divide.

5. $3\overline{)84}$ **6.** $6\overline{)72}$ **7.** $7\overline{)952}$ **8.** $6\overline{)84}$

9. $8\overline{)96}$ **10.** $3\overline{)933}$ **11.** $4\overline{)88}$ **12.** $5\overline{)80}$

13. $4\overline{)716}$ **14.** $2\overline{)684}$ **15.** $3\overline{)54}$ **16.** $5\overline{)65}$

17. 575 ÷ 5 =_____ **18.** 84 ÷ 7 =_____ **19.** 940 ÷ 2 =_____

20. 66 ÷ 3 =_____ **21.** 628 ÷ 4 =_____ **22.** 99 ÷ 9 =_____

Solving Problems

Solve.

23. Mel, Karen, Larry, and Mary Jean bowled three games in their Wednesday night bowling league. Divide each total score by 3 to find each bowler's average. Complete the score sheet.

NAME	TOTAL SCORE	AVERAGE
Mel	375	
Karen	381	
Larry	342	
Mary Jean	366	

Check your answers on pages 82-83.

19 One Digit Division With Remainders

Four friends shared the expenses of a fishing trip. The cost of the gas was \$125. How much money did each person owe for gas? What was the remainder?

Think

To find out how much each person owed for gas, divide the cost of gas by the number of people sharing the cost. The value left over is called the *remainder*. What you do with the remainder is determined by the situation.

Do

Step 1. Place the number in a division bracket.

$$\begin{array}{r} \text{- -} \\ 4\overline{)\$125} \end{array}$$

Step 2. Determine where the quotient will begin.

Since 1 cannot be divided by 4, move to the next digit. 12 can be divided by 4. The first digit of the quotient should be placed above the 2.

Step 3. Use these steps to divide.

Divide. $12 \div 4 = 3$

Multiply. $3 \times 4 = 12$

Subtract. $12 - 12 = 0$

Bring down the next digit, 5.

$$\begin{array}{r} 3 \\ 4\overline{)\$125} \\ -\underline{12} \\ 05 \end{array}$$

Step 4. Repeat the process.

Divide. $5 \div 4 = 1$

Multiply. $1 \times 4 = 4$

Subtract. $5 - 4 = 1$

$$\begin{array}{r} \$\ 31 \\ 4\overline{)\$125} \\ -\underline{12} \\ 05 \\ \underline{4} \\ 1 \end{array}$$

Step 5. The value left over is called the remainder.

Write "R 1" next to the quotient.

Each friend owed \$31 for gas. There was a \$1 remainder.

Try These

Divide. Place the number in a bracket if necessary.

1. $5\overline{)73}$ **2.** $7\overline{)131}$ **3.** 49 ÷ 6 =_____ **4.** 517 ÷ 4 =_____

PRACTICE

Divide.

5. $5\overline{)228}$ **6.** $8\overline{)89}$ **7.** $2\overline{)487}$ **8.** $8\overline{)459}$

9. $7\overline{)498}$ **10.** $6\overline{)83}$ **11.** $5\overline{)487}$ **12.** $7\overline{)148}$

13. 317 ÷ 7 =_____ **14.** 412 ÷ 5 =_____ **15.** 563 ÷ 2 =_____

16. 29 ÷ 6 =_____ **17.** 921 ÷ 8 =_____ **18.** 211 ÷ 3 =_____

Solving Problems

Solve.

19. It is 826 miles to Diamond Lake. Driving was shared equally among three of the friends. How many miles did each person drive?

____

20. Two of the friends decided to split the cost of a used tent and camp stove. The total price was $115. What amount did each pay?

____

21. Cost for the week's groceries at Diamond Lake was $87. If they split the cost evenly, what part of the grocery bill did each of the four friends pay?

____

Check your answers on page 83.

More About Dividing and Checking

Virginia needs to schedule vans to take the residents of Hampton Center Retirement Village on a trip. Each van can take eight passengers. How many vans are needed to transport 93 people?

Think

To find how many vans are needed to transport 93 people, divide the number of passengers by the number of seats in each van. Use multiplication to check for accuracy.

Do

Step 1. Place the number in a division bracket.

```
 - -
8)93
```

Step 2. Determine where the quotient will begin.
The first digit of the quotient will be placed above the 9.

Step 3. Use these steps to divide.

Divide. $9 \div 8 = 1$

Multiply. $8 \times 1 = 8$

Subtract. $9 - 8 = 1$

Bring down the next digit, 3.

```
  1
8)93
  8
  13
```

Step 4. Repeat the process.

Divide. $13 \div 8 = 1$

Multiply. $8 \times 1 = 8$

Subtract. $13 - 8 = 5$

5 is the remainder.

```
  11 R5
8)93
 - 8
   13
    8
    5
```

Check by multiplication.

Division

```
quotient ——> 11 R5
divisor ——> 8)93
            - 8
              13
               8
               5
```

Multiplication

Multiply the divisor by the quotient.
Add the remainder.

```
  11
 × 8
  88
 + 5
  93
```

The answer should be the number you started with.

Virginia needs to order 12 vans for the residents. Eleven vans will carry eight passengers each, and the twelfth van will carry the five remaining passengers.

Try These

Divide. Check by multiplication.

1. $4\overline{)78}$ *check*

2. $5\overline{)491}$ *check*

3. $4\overline{)585}$ *check*

4. $8\overline{)236}$ *check*

PRACTICE

Divide. Check by multiplication.

5. $7\overline{)52}$ *check*

6. $3\overline{)76}$ *check*

7. $5\overline{)412}$ *check*

8. $4\overline{)467}$ *check*

9. $6\overline{)154}$ *check*

10. $9\overline{)381}$ *check*

11. $3\overline{)2,737}$ *check*

12. $6\overline{)6,584}$ *check*

13. $9\overline{)4,286}$ *check*

Solving Problems

Solve. Check your answers by multiplication.

14. Thi Vinh needs to drive 3,421 miles in four days. If she divides the driving distance equally, how far should she drive in one day?

____

15. Kenny sorts laundry at the Broadway Linen and Uniform Company. How many 8-pound piles of laundry can he make from 237 pounds?

____

16. Members of the Reading Club were assigned to bind used books into sets of four for the annual book sale. How many sets of four can be made from 1,522 books?

____

Check your answers on pages 84-85.

Lou wants to hire part-time employees to work in his new Stop 'n' Shop. The store will be open 125 hours a week. Lou needs to have one part-time worker at the store at all times. How many employees does he need to hire if each is scheduled to work 25 hours a week?

Think

To find how many workers Lou needs to hire, divide the number of hours he plans to be open by the number of hours he wants each employee to work.

Do

Step 1. Place the number in a division bracket. $25\overline{)125}$

Step 2. Determine where the quotient will begin.

Since 1 cannot be divided by 25, move to the next digit.

Since 12 cannot be divided by 25 either, move to the next digit.

125 can be divided by 25. The quotient begins above the 5.

Step 3. Use these steps to divide.

$$\begin{array}{r} 6 \\ 25\overline{)125} \\ -\,150 \end{array}$$

Divide.

$125 \div 25 =$

To estimate how many times one number goes into another, divide using the leading digits.

Divide 12 by 2. $12 \div 2 = 6$

Try 6 as the quotient.

Multiply.

$6 \times 25 = 150$

Since 150 is too large, decrease the value of the quotient to 5.

Multiply.

$5 \times 25 = 125$ (That number works!)

Subtract 125 − 125 = 0

$$\begin{array}{r} 5 \\ 25\overline{)125} \\ -\underline{125} \\ 0 \end{array}$$

Lou will need to hire 5 part-time workers.

Try These

Divide.

1. $36\overline{)72}$ 2. $53\overline{)583}$ 3. $26\overline{)338}$ 4. $31\overline{)837}$

PRACTICE

Divide.

5. $13\overline{)1{,}586}$ 6. $24\overline{)912}$ 7. $28\overline{)896}$ 8. $49\overline{)882}$

9. $22\overline{)1{,}958}$ 10. $45\overline{)1{,}665}$ 11. $63\overline{)945}$ 12. $16\overline{)384}$

13. $33\overline{)792}$ 14. $45\overline{)1{,}575}$ 15. $82\overline{)2{,}952}$ 16. $64\overline{)3{,}264}$

17. $39\overline{)1{,}053}$ 18. $48\overline{)384}$ 19. $51\overline{)459}$ 20. $67\overline{)5{,}427}$

Solving Problems

Solve.

21. The display cases for candy and snacks are 15 feet long. How many display cases can Lou set up in a 75-foot area?

____

22. An employee has 336 bottles of vitamins to put in a display. Each display column holds 12 bottles of vitamins. How many columns will the employee need to display the bottles?

____

23. Lou wants to distribute store coupons at his grand opening. Lou can fit 12 coupons on a sheet of paper. How many copies of the page will he need to make to get 408 coupons?

____

Check your answers on page 85.

Ronnie stood in a line behind 98 people to ride the whirly-bird at the state fair. A group of 24 people can ride the whirly-bird at each stop. How many groups will Ronnie need to wait for before he gets on the ride?

Think

Group is a key word that tells you to divide. To find the number of groups Ronnie needs to wait for, divide the number of people waiting in line by 24.

Do

Step 1. Place the number in a division bracket. $24\overline{)98}$

Step 2. Determine where the quotient will begin.

Since 9 cannot be divided by 24, move to the next digit. 98 can be divided by 24.

The answer begins above the 8.

Step 3. Use these steps to divide.

Divide. $98 \div 24 =$

Estimate how many times 24 goes into 98 using the leading digits.

Divide 9 by 2. $9 \div 2 = 4$

Try 4 as the quotient.

Multiply. $4 \times 24 = 96$ (That number works!)

Subtract. $98 - 96 = 2$

2 is the remainder.

$$\begin{array}{r} 4\ R2 \\ 24\overline{)98} \\ -\underline{96} \\ 2 \end{array}$$

Ronnie had to wait for 4 groups to ride the whirly-bird. Since he was the second person in line after the fourth group got on the ride, he was in the fifth group.

Try These

Divide.

1. $32\overline{)78}$ **2.** $41\overline{)359}$ **3.** $60\overline{)912}$ **4.** $75\overline{)461}$

PRACTICE

Divide.

5. $47\overline{)83}$ **6.** $33\overline{)563}$ **7.** $52\overline{)189}$ **8.** $16\overline{)742}$

9. $85\overline{)104}$ **10.** $30\overline{)51}$ **11.** $62\overline{)332}$ **12.** $41\overline{)769}$

13. $29\overline{)506}$ **14.** $55\overline{)391}$ **15.** $46\overline{)196}$ **16.** $82\overline{)287}$

17. $13\overline{)74}$ **18.** $24\overline{)482}$ **19.** $68\overline{)923}$ **20.** $57\overline{)644}$

Solving Problems

Solve.

21. The State Fair Officials want to hang banners above the display areas at the fair. Each banner will be 28 feet long. How many banners can be cut from a roll of plastic that is 470 feet long?

____

22. How many groups of 12 carnival midway tickets can be sold from a roll of 157?

____

23. The hot dog vendor is preparing 225 hot dogs. How many packages of hot dog buns, 24 to a package, will she need?

____

Check your answers on pages 85-86.

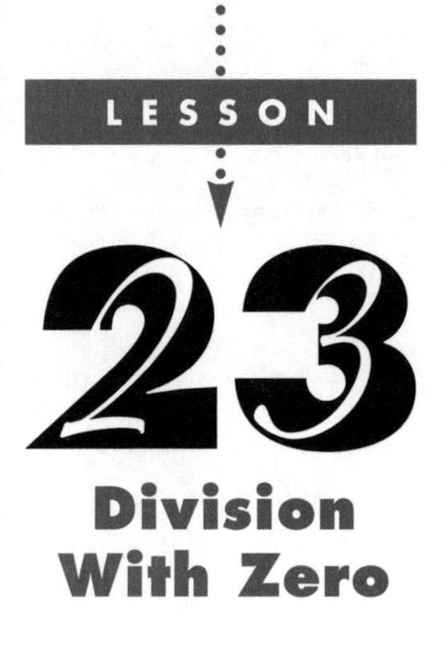

The Macias plan to have 100 people at their daughter's wedding reception. They will borrow tables that will each seat 8 people for the backyard reception. How many tables do they need?

Think

To find the number of tables needed, divide the number of people attending the reception by the number of people who can sit at one table. Zeros in the problem do not change the process of dividing.

Do

Step 1. Place the number in a division bracket.

Step 2. Determine where the quotient will begin. The answer begins above the first 0.

$$8\overline{)100}$$

Step 3. Use these steps to divide.
Divide. $10 \div 8 = 1$
Multiply. $1 \times 8 = 8$
Subtract. $10 - 8 = 2$
Bring down the next digit, 0.

$$\begin{array}{r} 1 \\ 8\overline{)100} \\ -\underline{8} \\ 20 \end{array}$$

Step 4. Repeat the process.
Divide. $20 \div 8 = 2$
Multiply. $2 \times 8 = 16$
Subtract. $20 - 16 = 4$
4 is the remainder.

$$\begin{array}{r} 12R4 \\ 8\overline{)100} \\ -\underline{8} \\ 20 \\ -\underline{16} \\ 4 \end{array}$$

The Macias need to borrow 13 tables for the reception because an extra table is needed for the 4 remaining people.

Try These

Divide.

1. $5\overline{)330}$ **2.** $4\overline{)607}$ **3.** $7\overline{)4{,}000}$ **4.** $3\overline{)5{,}060}$

PRACTICE

Divide.

5. $3\overline{)503}$ **6.** $6\overline{)8{,}008}$ **7.** $5\overline{)308}$ **8.** $9\overline{)640}$

9. $6\overline{)9{,}200}$ **10.** $7\overline{)400}$ **11.** $8\overline{)1{,}206}$ **12.** $3\overline{)8{,}091}$

13. $6\overline{)702}$ **14.** $5\overline{)2{,}058}$ **15.** $3\overline{)3{,}003}$ **16.** $8\overline{)60}$

Solving Problems

Solve.

17. Kamia bought 50 yards of fabric to make table covers. Each table cover requires 3 yards of fabric. How many table covers can she make from the fabric?

____

18. Relatives are driving 550 miles to attend the wedding. If they drive 55 miles an hour, how many hours will it take them to get to the wedding?

____

19. Five families shared the cost equally of a $70 wedding gift to the newlyweds. How much did each family pay toward the gift?

____

Check your answers on page 86.

Division By Numbers Ending in Zero

Mr. Landis runs a small market. He stops by the bank every morning to get cash for the day's transactions. Mr. Landis wants $400 worth of $20 bills. How many $20 bills will the bank teller give him for that amount of money?

Think

To find how many $20 bills the teller will give Mr. Landis, divide $400 by $20. The zeros in the problem do not change the process of dividing.

Do

Step 1. Place the number in a division bracket. $20\overline{)400}$

Step 2. Determine where the quotient will begin.

Since 20 cannot be divided into 4, move to the next digit.

40 can be divided by 20. The quotient will begin above the first zero.

Step 3. Use these steps to divide.

Divide. $40 \div 20 = 2$

Multiply. $2 \times 20 = 40$

Subtract. $40 - 40 = 0$

There is no remainder, but there is still another digit to bring down.

$$\begin{array}{r} 20 \\ 20\overline{)400} \\ -\underline{40} \\ 0 \end{array}$$

Step 4. Write a zero in the quotient above the final zero in 400.

The bank teller will give Mr. Landis twenty $20 bills.

Try These

Divide.

1. $10\overline{)40}$ **2.** $60\overline{)600}$ **3.** $40\overline{)800}$ **4.** $300\overline{)9,900}$

Divide.

5. $50\overline{)300}$ 6. $30\overline{)2,100}$ 7. $10\overline{)580}$ 8. $70\overline{)490}$

9. $20\overline{)820}$ 10. $80\overline{)160}$ 11. $200\overline{)4,400}$ 12. $100\overline{)6,000}$

13. $20\overline{)206}$ 14. $30\overline{)429}$ 15. $60\overline{)517}$ 16. $10\overline{)708}$

17. $50\overline{)331}$ 18. $400\overline{)842}$ 19. $300\overline{)6,611}$ 20. $70\overline{)9,523}$

Solving Problems

Solve.

21. Windmill Nursery donated 2,300 tulip bulbs to a city beautification project. The bulbs were packaged in bags of 50. How many packages of bulbs were donated?

____

22. Garbage bags are sold in bundles of 100. The project committee needs 1,000 bags. How many bundles should they buy?

____

23. Forty volunteers worked a total of 1,200 hours on the city project. If the volunteers worked the same number of hours, how many hours did each work on the project?

____

Check your answers on page 86.

Rounding and Estimating in Division

The Cub Scouts are planning a pancake breakfast to raise money. They estimate they will need to make 984 pancakes for the event. One bag of "Mr. Flapjack" Pancake Mix makes about 82 pancakes. Approximately how many bags will they need?

Think

Approximately is a key word that tells you to estimate. To estimate how many bags of "Mr. Flapjack" will be needed, round 984 and 82 to their leading digits and then divide.

Do

Step 1. Place the number in a division bracket.

$$82\overline{)984} \longrightarrow 80\overline{)1{,}000}$$

Step 2. Round each number to the leading digit.

Step 3. Determine where the quotient will begin. The quotient begins above the second 0.

Step 4. Divide using the rounded numbers.

$$\begin{array}{r} 12 \\ 80\overline{)1{,}000} \\ -\underline{80} \\ 200 \\ -\underline{160} \\ 40 \end{array}$$

Step 5. Drop any remainder.
The scouts will need approximately 12 bags of "Mr. Flapjack."

Try These

Round and estimate.

round

1. $37\overline{)4{,}102}$

2. $142\overline{)286}$

Round and estimate. Then divide for the exact answer. Make sure your answer is reasonable.

3. $22\overline{)752}$

4. $68\overline{)4{,}029}$

Round and estimate.

5. $53\overline{)224}$ **6.** $49\overline{)841}$ **7.** $12\overline{)163}$

8. $55\overline{)977}$ **9.** $16\overline{)761}$ **10.** $91\overline{)1{,}845}$

11. $112\overline{)1{,}390}$ **12.** $247\overline{)8{,}006}$ **13.** $486\overline{)1{,}660}$

Round and estimate. Then divide for an exact answer. Make sure your answer is reasonable.

14. $12\overline{)493}$ **15.** $25\overline{)760}$

Solving Problems

Solve.

16. Roberto types 34 words a minute on the computer. About how long will it take him to type a 590-word article for a newsletter?

____

17. A custodial staff cleans 8,640 square feet of office floor space. A can of floor wax covers about 720 square feet. Approximately how many cans of floor wax does the staff need to clean the office?

____

Check your answers on pages 86-87.

LESSON

Division Review

Divide.

1. $35 \div 7 =$ _____

2. $16 \div 4 =$ _____

3. $138 \div 6 =$ _____

4. $471 \div 3 =$ _____

5. $5\overline{)45}$

6. $9\overline{)72}$

7. $8\overline{)64}$

8. $2\overline{)26}$

9. $2\overline{)148}$

10. $9\overline{)333}$

11. $8\overline{)984}$

12. $5\overline{)260}$

13. $27\overline{)891}$

14. $5\overline{)1{,}748}$

15. $7\overline{)3{,}762}$

16. $55\overline{)115}$

17. $5\overline{)300}$

18. $8\overline{)4{,}006}$

19. $6\overline{)1{,}050}$

20. $75\overline{)304}$

21. $70\overline{)1{,}472}$

22. $60\overline{)5{,}935}$

23. $400\overline{)1{,}600}$

24. $30\overline{)6{,}204}$

Divide. Check by multiplication.

check

25. $6\overline{)522}$

26. $7\overline{)655}$

27. $83\overline{)1{,}012}$

28. $54\overline{)460}$

Round and estimate. Then divide for an exact answer. Make sure your answer is reasonable.

round

29. $31\overline{)182}$

30. $49\overline{)523}$

31. $300\overline{)9{,}480}$

32. $518\overline{)4{,}686}$

33. $23\overline{)967}$

34. $64\overline{)2{,}518}$

Solving Problems

Solve.

35. Felipe canned 413 jars of salsa. He can store 18 jars on one shelf. How many shelves does he need to store all the salsa?

____

36. The 15-member paint crew split up the painting job at Municipal Stadium. There are 8,445 seats in the stadium. How many seats did each crew member paint?

____

37. On an assembly line Marla snaps about 1,800 plastic parts together in a 40-hour work week. Estimate the number of plastic parts Marla assembles in one hour.

____

38. How many weeks will it take Jake to save $95 if he saves $5 a week?

____

39. Marie and Eileen are roommates in an apartment that rents for $380 a month. What is each roommate's share of the rent? What would each person's rent be if they shared the apartment with another friend?

____

40. Fifty-six people turned out for a block party. They decided to break into 4 teams to play volleyball. How many people were on each team?

____

Check your answers on pages 87-88.

LESSON

Multiplication and Division Review

Multiply or divide.

1. $37 \times 8 =$ _____
2. $210 \div 3 =$ _____
3. $306 \div 9 =$ _____
4. $95 \times 7 =$ _____
5. $16 \times 48 =$ _____
6. $612 \div 4 =$ _____

7. $\begin{array}{r} 4{,}387 \\ \times\ 9 \\ \hline \end{array}$

8. $3\overline{)386}$

9. $9\overline{)468}$

10. $\begin{array}{r} 623 \\ \times\ 80 \\ \hline \end{array}$

11. $61\overline{)2{,}684}$

12. $52\overline{)1{,}161}$

13. $\begin{array}{r} 407 \\ \times\ 78 \\ \hline \end{array}$

14. $20\overline{)628}$

15. $33\overline{)295}$

16. $\begin{array}{r} 426 \\ \times\ 83 \\ \hline \end{array}$

17. $\begin{array}{r} 7{,}105 \\ \times\ 671 \\ \hline \end{array}$

18. $4\overline{)8{,}265}$

19. $\begin{array}{r} 416 \\ \times\ 200 \\ \hline \end{array}$

20. $\begin{array}{r} 6{,}500 \\ \times\ 379 \\ \hline \end{array}$

21. $38\overline{)4{,}600}$

22. $\begin{array}{r} 36 \\ \times\ 100 \\ \hline \end{array}$

Round and estimate.

23. $\begin{array}{r} 471 \rightarrow 500 \\ \times\ 65 \rightarrow \ \ 70 \end{array}$

24. $57\overline{)6{,}324}$

25. $9\overline{)888}$

26. $\begin{array}{r} 932 \\ \times\ 456 \\ \hline \end{array}$

27. $\begin{array}{r} 9{,}215 \\ \times\ 18 \\ \hline \end{array}$

28. $210\overline{)2{,}340}$

Round and estimate. Then multiply or divide for an exact answer. Compare the exact answer to the estimate. Make sure your answer is reasonable.

29. $\begin{array}{r} 287 \\ \times\ 403 \\ \hline \end{array}$

30. $67\overline{)7{,}821}$

31. $23\overline{)9{,}931}$

32. $\begin{array}{r} 3{,}860 \\ \times\ 57 \\ \hline \end{array}$

33. $\begin{array}{r} 9{,}602 \\ \times\ 78 \\ \hline \end{array}$

34. $\begin{array}{r} 871 \\ \times\ 35 \\ \hline \end{array}$

Check your answers on page 88.

Solving Problems With Multiplication and Division

A cross-country team held a "Fun Run" to raise money. Each of the 22 members of the team raised $25. How much money did the entire team raise?

Think

Using a problem-solving chart can help solve problems. The steps on the chart are:
Find the specific question the problem is asking.
Locate all the information you need to solve the problem.
Choose a method. Look for key words.
Work the problem using the information and the method.
Answer the question. (THINK: Is the answer reasonable?)

Do

Complete each step. Fill out the chart.
Question: Find the specific *question* the problem is asking.

Question	Information	Method	Process	Answer
How much money did the entire team raise?				

Information: Find all the *information* you need to solve the problem.

Question	Information	Method	Process	Answer
How much money did the entire team raise?	22 members raised $25 each			

Method: Look for key words that tell you what operation to use. Sometimes the *method* may involve more than one operation. *Entire* is a key word that tells you to multiply.

Question	Information	Method	Process	Answer
How much money did the entire team raise?	22 members raised $25 each	Multiply		

Process: Work out the problem using your *information* and *method.*

Question	Information	Method	Process	Answer
How much money did the entire team raise?	22 members raised $25 each	Multiply	22 × $25 110 440 $550	

Answer: Answer the question. Think, is the answer reasonable?

Question	Information	Method	Process	Answer
How much money did the entire team raise?	22 members raised $25 each	Multiply	22 × $25 110 440 $550	The team raised $550.

Yes, the amount is reasonable.

Try These

Use the problem-solving chart to help you solve these problems.

1. To train for the cross-country team, the coach advised Christine and Michelle to run 300 miles in 3 months. How many miles should they run in 1 month?

Question	Information	Method	Process	Answer

2. Michelle can run a mile in 8 minutes. How long will it take her to run 3 miles?

Question	Information	Method	Process	Answer

PRACTICE

Use the problem-solving chart to solve these problems.

3. On Saturday morning Christine ran for 100 minutes. At a pace of about 11 minutes a mile, about how many miles did she run?

Question	Information	Method	Process	Answer

4. Christine subscribed to *Runner* magazine. She paid $24 for 12 issues. What price did she pay per issue?

Question	Information	Method	Process	Answer

5. Michelle decided to run on an indoor track during bad weather. Fourteen laps on the track equals one mile. How many laps did she run for three miles?

Question	Information	Method	Process	Answer

Solving Problems

Solve these problems using the steps from the problem-solving chart. You may draw the chart if you like.

6. During the 1972 Olympics, a runner ran the 1,500-meter run in four minutes. If the runner kept an even pace during the run, how many meters did she run per minute?

____

7. City officials are planning a marathon. They plan to set up bleachers for spectators every two miles along the course, including a set of bleachers at the finish line. If a marathon is 26 miles in length, how many sets of bleachers will the officials need?

____

8. Rakim is a distance runner. He runs an average of 310 meters per minute. At that pace, how many meters can he run in 30 minutes?

____

Check your answers on pages 88-89.

LESSON

Problem-Solving Review

Use the problem solving chart to help you solve the following problems.

1. During a 12-day heat wave about 5,800 people swam at Fairview City Pool. Estimate the number of people who swam at the pool each day.

Question	Information	Method	Process	Answer

2. Sameer's son invited 5 boys to Fantasy Fun Park for his birthday. Admission to the park was $10 per child. Sameer used three "2 for the price of 1" discount coupons. What was the cost of the six boys' admission?

Question	Information	Method	Process	Answer

3. Ralph bought an engagement ring for $436. He made a down payment of $100 and will make 12 monthly payments to pay off the ring. How much will the payments be?

Question	Information	Method	Process	Answer

4. Woodrow Day Care Center charges $90 a week for child care. The second child in the family is charged $75 a week. Each additional child is charged $35 a week. What is the weekly cost for a family with three children?

Question	Information	Method	Process	Answer

5. Ellen brought 68 handmade baskets to the craft show. She arranged 7 baskets each on 8 shelves. How many baskets did Ellen have left over?

Question	Information	Method	Process	Answer

Solve these problems using the steps from the problem solving chart.

6. Lynette earns a monthly check of $956 after taxes. Family insurance of $78 and union dues of $112 are then deducted. What is her monthly take home pay?

Question	Information	Method	Process	Answer

7. A school needs 25 jump ropes. To save money, the school decides to buy clothesline rope and cut it into 4-foot lengths. How many feet of clothesline rope will the school need to buy to make the jump ropes?

Question	Information	Method	Process	Answer

8. Thuc needs to complete 300 hours of aerobic exercise to achieve a fitness award. He completed 78 hours of exercise during the first session, 102 hours during the second, and 49 hours during the third session. How many more hours of aerobic exercise does he need to complete in order to get the award?

Question	Information	Method	Process	Answer

9. Cathy has been saving money to buy a used sewing machine. She already has $200 in savings. The sewing machine she needs costs $525. If Cathy can save $30 each month, how many months will it take her to save enough to buy the sewing machine?

Question	Information	Method	Process	Answer

10. Six friends decide to buy an old car and rebuild it. They can each put in $75 towards the cost of the car. The car they want will cost them $600. How much more do they need to buy the car?

Question	Information	Method	Process	Answer

Check your answers on pages 89-90.

Lacina wants to put a wire fence around her vegetable garden. The shape of the garden is square. The length of each side is 10 feet. How much wire fence does she need?

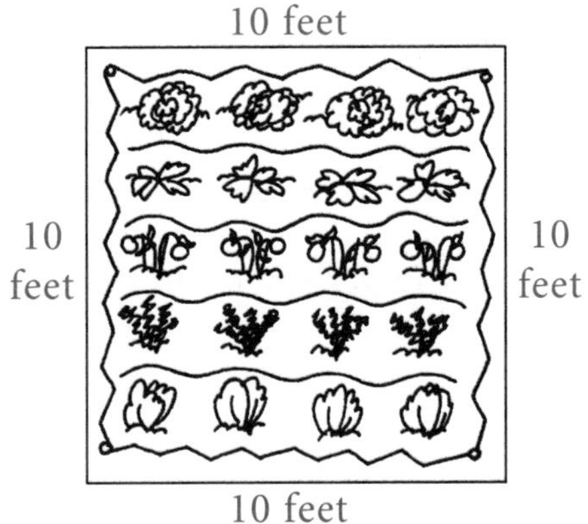

Think

The distance around a shape is called the *perimeter*. To find the perimeter, add the length of all the sides.

Do

Step 1. Find the length of the sides of the square.

Since the garden is a square, all the sides are the same length.

Each side measures 10 feet.

$$\begin{array}{r} \mathbf{10} \\ \mathbf{10} \\ \mathbf{10} \\ \underline{\mathbf{+10}} \\ 40 \end{array}$$

Step 2. Add the lengths of all the sides.

NOTE: Because the sides are the same length, you could also solve the problem by multiplying.

$10 \times 4 = 40$

Lacina needs 40 feet of wire fence for her garden.

Try These

Find the perimeter of each.

1. ____ **2.** ____

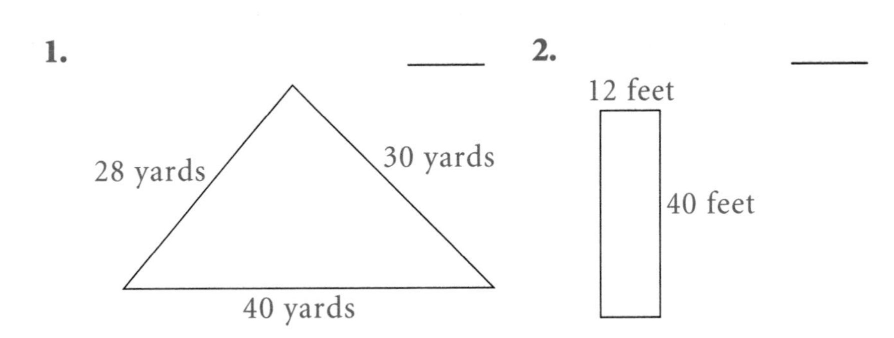

3. a frame for a rectangular picture which is 8 inches by 12 inches ______

4. a fringe for the edge of a table cloth that is 60 inches by 80 inches ______

PRACTICE

Find the perimeter of each.

5. ______

4 miles

2 miles

6. ______

35 yards

15 yards

37 yards

7. ______

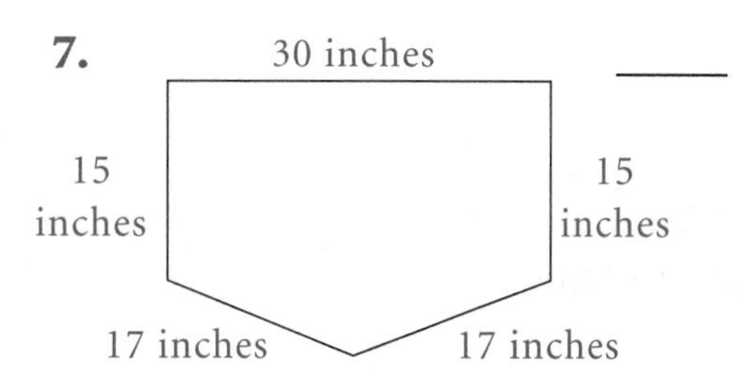

8. ______

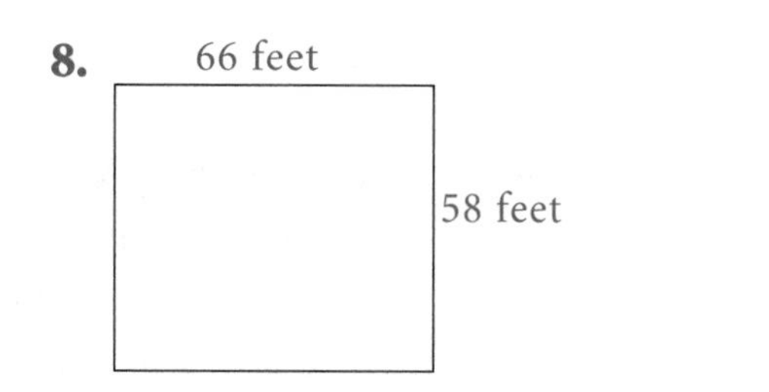

9. a wallpaper border for a rectangular bedroom which measures 15 feet by 10 feet ______

10. a triangle sand box that is 5 feet by 7 feet by 9 feet ______

Solving Problems

Solve.

11. Daniel is finishing his basement. He wants to put a baseboard around a room that is 14 feet on two sides, 12 feet on one side, and 9 feet on the other side. How much baseboard does he need?

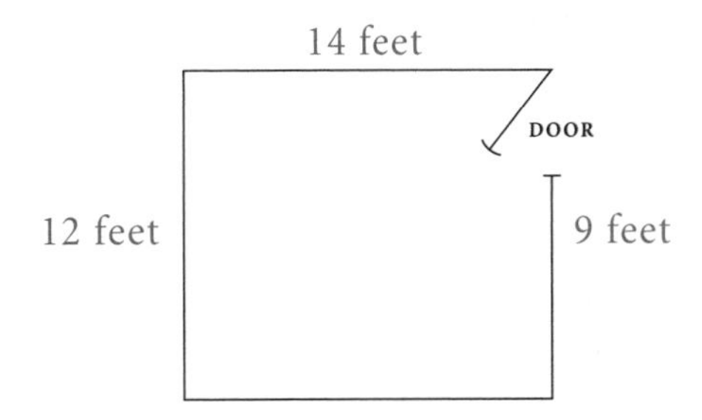

____

12. The sections of the path around the lake are 12,875 feet, 9,500 feet, 7,130 feet, 900 feet, and 6,910 feet. What is the distance of the path around the lake?

____

13. Kara wants to build a 9 foot by 4 foot dog run for her dalmatian. She will use the house as one of the 9-foot sides. How many feet of fencing does she need for the remaining three sides?

____

Check your answers on page 90.

Bill and his children are working on a backyard patio project. Bill figured out that it would take 36 tiles to cover the length of the patio and 28 tiles to cover the width. How many tiles will Bill and his children need to fill the inside area of the patio?

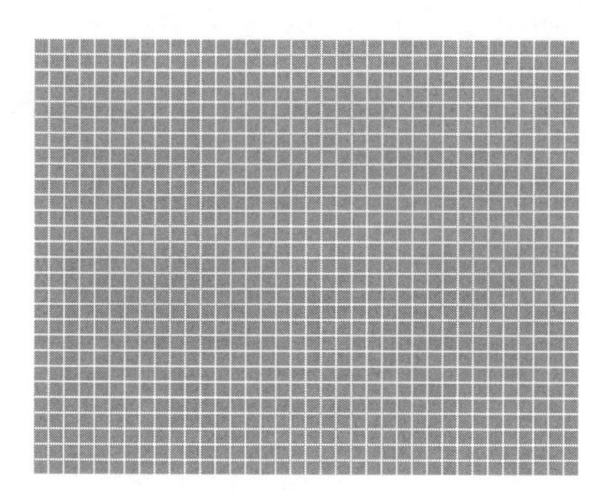

Think

The inside surface of a shape is called the *area.* Area is measured in "square" units. To find the area, multiply the length times the width.

Do

Step 1. Find the length and the width of the sides.
The length of the patio is 36 tiles.
The width of the patio is 28 tiles.

```
   23 tiles
×  28 tiles
  288
+ 720
1,008 square tiles
```

Step 2. Multiply the length times the width.

Step 3. Label the answer in square units.
Since the patio is measured in tiles, the area measures 1,008 square tiles.

They will need 1,008 tiles to fill the inside area of the patio.

Try These

Find the area of each.

1. 9 feet ______

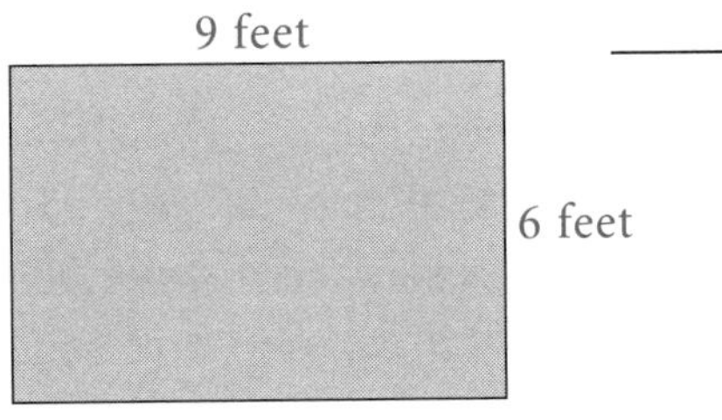

6 feet

2. 4 miles ______

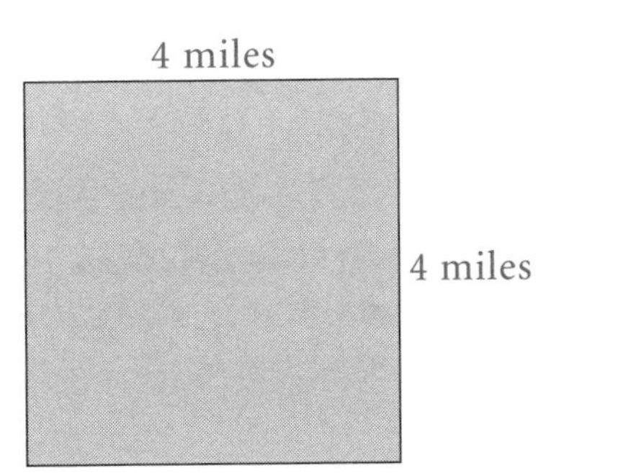

4 miles

3. a rectangle that measures 34 feet by 17 feet ______

4. carpet for a 9 foot by 12 foot den ______

PRACTICE

Find the area of each.

5. 120 feet ______

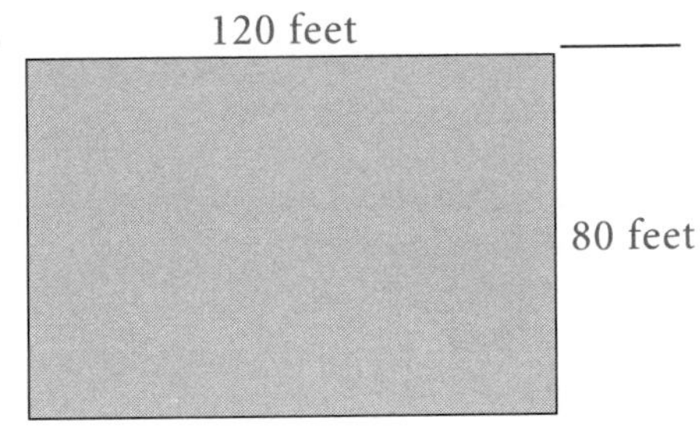

80 feet

6. ______

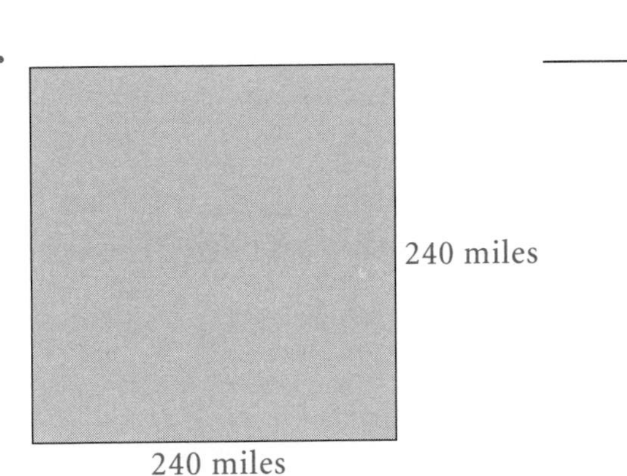

240 miles

240 miles

7. 22 yards ______

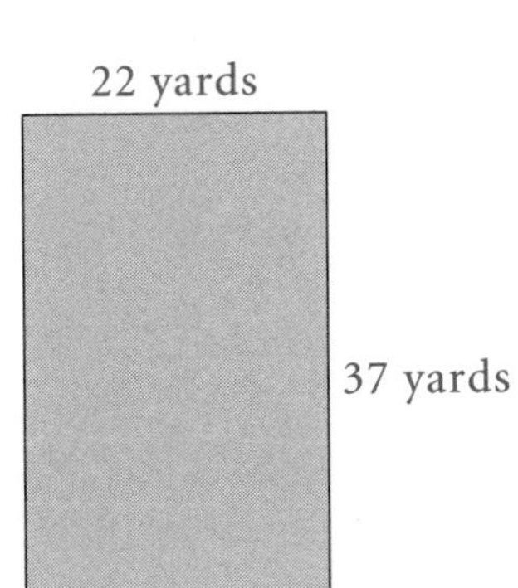

37 yards

8. 44 inches ______

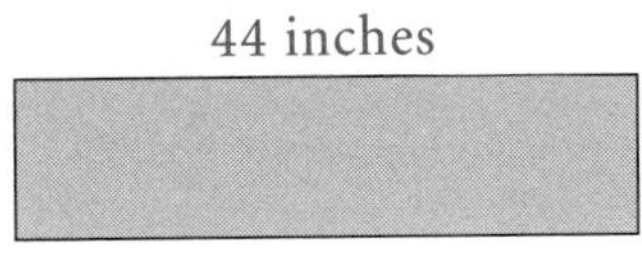

11 inches

9. screen for a door with a frame of 3 feet by 7 feet ______

10. a cover for a square-shaped hot tub which is 5 feet on each side ______

Solve.

11. The new football field at Valley Junior College will be covered with artificial turf. How much turf is needed for an area that is 130 yards long and 70 yards wide?

____

12. Samira is seeding a yard. The yard measures 35 feet by 50 feet. One bag of grass seed covers a 1,500-square-foot area. Will one bag of grass seed be enough to cover the yard?

____

13. A 10-foot by 25-foot section of Big Bob's Burger Drive-Thru needs concrete repair. What is the area of that section?

____

Check your answers on page 91.

LESSON

Averages

Bea bowled three games in her Friday-morning league. The first game she bowled 125. The second game she bowled 137. The third game she bowled 107. What was her average score?

Think

The typical value of a group of numbers is called the average. To find the average, add the group of numbers. Then divide the total by the number of numbers in the group.

Do

Step 1. Find the total number of points in all three games.

$$\begin{array}{r} 125 \\ 137 \\ +\ 107 \\ \hline 369 \end{array}$$

Step 2. Divide the total by the number of games bowled.

$$\begin{array}{r} 123 \\ 3\overline{)369} \\ -\ 3 \\ \hline 06 \\ -\ 6 \\ \hline 09 \\ -\ 9 \\ \hline 0 \end{array}$$

Bea's average score was 123.
NOTE: Since the division step is used in figuring the average, the average may include a remainder. Each situation determines how the remainder is used.

Try These

Find the average.

1. 15, 12, 18, 13, 17 _____ **2.** $10, $8, $12, $10 _____

3. What is the average of the test scores? 81, 85, 95, 77, 87 _____

4. What is the average weight of the linemen on the team? 250 lbs, 300 lbs, 305 lbs, 295 lbs, 310 lbs, 298 lbs _____

PRACTICE

Find the average.

5. 4, 8, 12, 6, 5 _____

6. 21, 422, 21, 224, 21, 422, 21, 204 _____

7. $51, $37, $42, $50 _____

8. 223, 224, 225, 227, 228, 229 _____

9. What is the average high temperature? 98°, 99°, 103°, 100°, 96°, 96° _____

10. David lost a total of 40 pounds in three months. What was his average monthly weight loss? _____

Solving Problems

Solve.

11. What is the average age of students in the advanced Spanish class at Metro Community School? The ages of the students are: 40, 25, 38, 44, 50, and 19.

_____

12. Katie delivered 12 Sing-a-Grams on Valentine's Day. She earned $60 in tips. How much in tips did she average per delivery?

_____

13. The top six drivers and their times are posted on the chart. What was the average finish time for the top three drivers?
(HINT: Top drivers show lowest times.)

Johnston County Motor Track

Name	Time(seconds)
Gradowski	30
Talcott	28
Rulo	23
Baiham	24
Plaza	25
Burke	26

_____

Check your answers on page 91.

Answer Key

Lesson 1 ······▶ WHAT IS MULTIPLICATION? (PAGE 3)

1. $5 + 5 + 5 = 15 \quad 5 \times 3 = 15$

2. $2 + 2 + 2 + = 8 \quad 2 \times 4 = 8$

3. $2 + 2 + 2 = 6 \quad 2 \times 3 = 6$

4. $4 + 4 + 4 + 4 = 16 \quad 4 \times 4 = 16$

5. $6 + 6 = 12 \quad 6 \times 2 = 12$

6. $9 + 9 + 9 + 9 = 36 \quad 9 \times 4 = 36$

7. $4 \times 9 = 36 \quad 9 \times 4 = 36$

8. $2 \times 8 = 16 \quad 8 \times 2 = 16$

9. $5 \times 7 = 35 \quad 7 \times 5 = 35$

10. $6 \times 9 = 54 \quad 9 \times 6 = 54$

11. $0 \times 2 = 0 \quad 2 \times 0 = 0$

12. $6 \times 3 = 18 \quad 3 \times 6 = 18$

13. $1 \times 3 = 3 \quad 3 \times 1 = 3$

14. $7 \times 8 = 56 \quad 8 \times 7 = 56$

15. $5 \times 6 = 30$ plants

16. $\$8 \times 2 = \$16$

Lesson 2 ······▶ MULTIPLICATION FACTS (PAGE 4–5)

1. $\begin{array}{r} 2 \\ \times 7 \\ \hline 14 \end{array} \qquad \begin{array}{r} 7 \\ \times 2 \\ \hline 14 \end{array}$

2. $\begin{array}{r} 4 \\ \times 8 \\ \hline 32 \end{array} \qquad \begin{array}{r} 8 \\ \times 4 \\ \hline 32 \end{array}$

3. $3 \times 5 = 15 \quad 5 \times 3 = 15$

4. $9 \times 6 = 54 \quad 6 \times 9 = 54$

5.	4	7	0	3	5	1	9	8	6	2
	× 0	× 0	× 0	× 0	× 0	× 0	× 0	× 0	× 0	× 0
	0	0	0	0	0	0	0	0	0	0
6.	5	2	6	4	0	8	7	3	9	1
	× 1	× 1	× 1	× 1	× 1	× 1	× 1	× 1	× 1	× 1
	5	2	6	4	0	8	7	3	9	1
7.	2	8	4	7	5	6	1	0	9	3
	× 2	× 2	× 2	× 2	× 2	× 2	× 2	× 2	× 2	× 2
	4	16	8	14	10	12	2	0	18	6
8.	4	6	5	2	9	7	1	3	8	0
	× 3	× 3	× 3	× 3	× 3	× 3	× 3	× 3	× 3	× 3
	12	18	15	6	27	21	3	9	24	0
9.	0	9	3	7	4	8	2	6	1	5
	× 4	× 4	× 4	× 4	× 4	× 4	× 4	× 4	× 4	× 4
	0	36	12	28	16	32	8	24	4	20
10.	7	4	6	0	1	9	8	5	3	2
	× 5	× 5	× 5	× 5	× 5	× 5	× 5	× 5	× 5	× 5
	35	20	30	0	5	45	40	25	15	10
11.	9	3	8	6	1	0	5	7	4	2
	× 6	× 6	× 6	× 6	× 6	× 6	× 6	× 6	× 6	× 6
	54	18	48	36	6	0	30	42	24	12
12.	3	5	2	0	6	8	7	9	1	4
	× 7	× 7	× 7	× 7	× 7	× 7	× 7	× 7	× 7	× 7
	21	35	14	0	42	56	49	63	7	28

13.	5	1	6	8	4	0	3	9	7	2
	$\times 8$	$\times 8$	$\times 8$	$\times 8$	$\times 8$	$\times 8$	$\times 8$	$\times 8$	$\times 8$	$\times 8$
	40	8	48	64	32	0	24	72	56	16

14.	2	7	0	6	5	8	1	3	4	9
	$\times 9$	$\times 9$	$\times 9$	$\times 9$	$\times 9$	$\times 9$	$\times 9$	$\times 9$	$\times 9$	$\times 9$
	18	63	0	54	45	72	9	27	36	81

15. $6 \times 1 = 6$ $1 \times 6 = 6$ $2 \times 3 = 6$ $3 \times 2 = 6$

16. $5 \times 9 = 45$ $9 \times 5 = 45$

17. $3 \times 3 = 9$ $9 \times 1 = 9$ $1 \times 9 = 9$

18. $5 \times 1 = 5$ $1 \times 5 = 5$

19. $2 \times 5 = 10$ $5 \times 2 = 10$

20. $3 \times 8 = 24$ $8 \times 3 = 24$ $4 \times 6 = 24$ $6 \times 4 = 24$

Lesson 3 ······► PLACE VALUE REVIEW (PAGES 6-7)

1. thousands

2. hundreds

3. tens

4. ones

5. thousands

6. hundred thousands

7. 47,863

8. 8,012,593

9. 159,362

ones	2
tens	6
hundreds	3
thousands	9
ten thousands	5
hundred thousands	1

10. 348,610

hundreds	6
ten thousands	4
ones	0
thousands	8
tens	1
hundred thousands	3

11. 7,385 = 7 thousands 3 hundreds 8 tens 5 ones

12. 4,911 = 4 thousands 9 hundreds 1 tens 1 ones

13. 840,712 b. eight hundred forty thousand, seven hundred twelve

14. 84,712 a. eighty-four thousand, seven hundred twelve

15. 804,127 c. eight hundred four thousand, one hundred twenty-seven

Lesson 4 ······➤ MULTIPLYING BY A ONE-DIGIT NUMBER (PAGE 9)

1. 159 2. 826 3. 305 4. 486 5. 96 6. 15 7. 166 8. 255 9. 183

10. 82 11. 36 12. 128 13. 248 14. 2,484 15. 1,266 16. 668 17. 126

18. 62 19. 66 20. 88 21. 1,688 22. 1,248 23. 108 24. 268 25. 1,263

26.	27.	28.
21 × 5 105 calories	12 × 4 $48 Yes.	122 × 4 $488

Lesson 5 ······➤ CARRYING IN ONE-DIGIT MULTIPLICATION (PAGES 10–11)

1.	2.	3.	4.	5.
2 24 × 5 120	1 613 × 4 2,452	1 72 × 6 432	413 × 2 826	6 18 × 8 144

6.	7.	8.	9.	10.
4 87 × 6 522	1 4 129 × 5 645	1 102 × 7 714	3 3 255 × 7 1,785	7 4 496 × 8 3,968

11.	12.	13.	14.	15.
1 3 538 × 4 2,152	1 32 × 6 192	3 96 × 5 480	2 64 × 6 384	4 6 747 × 9 6,723

16.	17.	18.	19.	20.
1 523 × 4 2,092	1 34 × 3 102	6 48 × 8 384	2 214 × 7 1,498	2 672 × 4 2,688

21.	22.	23.	24.	25.
2 5 139 × 6 834	5 2 763 × 9 6,867	14 $115 × 9 $1,035	2 25 × 4 100 appointments	2 53 × 7 371 newspapers

Lesson 6 ······➤ TWO-DIGIT MULTIPLICATION (PAGE 13)

1.	2.	3.	4.	5.	6.
23 × 31 23 + 690 713	421 × 42 842 16,840 17,682	32 × 13 96 320 416	121 × 23 363 2,420 2,783	32 × 31 32 960 992	413 × 32 826 12,390 13,216

7.	8.	9.	10.	11.	12.
24	111	124	432	332	42
× 12	× 36	× 22	× 13	× 23	× 21
48	666	248	1,296	996	42
240	3,330	2,480	4,320	6,640	840
288	3,996	2,728	5,616	7,636	882

13.	14.	15.	16.	17.	18.
131	222	13	423	34	141
× 32	× 41	× 21	× 33	× 12	× 21
262	222	13	1,269	68	141
3,930	8,880	260	12,690	340	2,820
4,192	9,102	273	13,959	408	2,961

19.	20.	21.
$24	33	312
× 12	× 12	× 42
48	66	624
240	330	12480
$288	396 blocks	13,104 seedlings

Lesson 7 ······► CARRYING IN TWO-DIGIT MULTIPLICATION (PAGE 15)

1. 4,200 **2.** 4,768 **3.** 1,406 **4.** 23,544 **5.** 2,106
6. 32,190 **7.** 792 **8.** 1,568 **9.** 288 **10.** 8,742
11. 23,856 **12.** 18,424 **13.** 3,149 **14.** 25,800 **15.** 9,212
16. 17,284 **17.** 2,100 **18.** 2,520 **19.** $ 1,072 **20.** 2,106 questions
21. 504 miles **22.** $4,465

Lesson 8 ······► CARRYING IN THREE-DIGIT MULTIPLICATION (PAGE 17)

1. 228,182 **2.** 376,090 **3.** 55,390 **4.** 153,406
5. 60,116 **6.** 268,598 **7.** 126,096 **8.** 327,684
9. 408,336 **10.** 217,056 **11.** 110,656 **12.** 364,041
13. 185,128 **14.** 205,933 **15.** 153,406 **16.** 334,696
17. 37,050 pair of shoes **18.** 29,325 customers **19.** 31,675 pizzas

Lesson 9 ······► MULTIPLYING WITH ZERO (PAGE 19)

1. 31,546 **2.** 224,084 **3.** 14,616 **4.** 38,722 **5.** 45,825 **6.** 307,020
7. 123,287 **8.** 49,910 **9.** 84,924 **10.** 83,076 **11.** 138,115 **12.** 140,805
13. 7,440 **14.** 20,474 **15.** 67,650 **16.** 270,225 **17.** 49,312 **18.** 71,440

19. $12
× 36
72
360
$432 more profit at $12 per hour

20. step 1
12
× 3
36
step 2
110
× 36
660
3300
3,960 food items

21. 150
× 25
750
3000
3,750 videos

Lesson 10 ······► MULTIPLYING BY 10, 100, AND 1000 (PAGE 21)

1. 4,900 **2.** 6,710 **3.** 152,000 **4.** 8,300 **5.** 920 **6.** 5,430

7. 210 **8.** 4,360 **9.** 18,400 **10.** 300 **11.** 156,200 **12.** 5,400

13. 873,000 **14.** 773,000 **15.** 60,000 **16.** 892,000 **17.** 59,600 **18.** 30

19. 96,000 **20.** 644,000 **21.** 8,750 **22.** 2,100

23. 250
× 1000
250,000 paper clips

24. 13
× $10
$130

25. 8
× 100
800 bushels

Lesson 11 ······► MULTIPLYING BY NUMBERS ENDING IN ZERO (PAGES 22–23)

1. 321
× 180
25680
32100
57,780

2. 42
× 20
840

3. 162
× 110
1620
16200
17,820

4. 38
× 500
19,000

5. 53
× 90
4,770

6. 126
× 40
5,040

7. 61
× 700
42,700

8. 307
× 50
15,350

9. 44
× 120
880
4400
5,280

10. 75
× 80
6,000

11. 28
× 130
840
2800
3,640

12. 98
× 60
5,880

13. 781
× 400
312,400

14. 87
× 50
4,350

15. 9
× 40
360

16. 33
× 20
660

17. 104
× 550
5200
52000
57,200

18. 868
× 900
781,200

19. 24
× 60
1,440 minutes

20. 50
× 30
1,500 30 days

21. 225
× 850
11250
180000
191,250 passengers

Lesson 12 ······▸ ROUNDING AND ESTIMATING IN MULTIPLICATION (PAGES 24–25)

round

1. 27 → 30
× 63 → 60
1,800

2. 593 → 600
× 141 → 100
60,000

3. 82 → 80
× 57 → 60
574 4,800
4,100
4,674

4. 2,483 → 2,000
× 367 → 400
17381 800,000
148980
744900
911,261

5. 409 → 400
× 51 → 50
20,000

6. 829 → 800
× 72 → 70
56,000

7. 38 → 40
× 67 → 70
2,800

8. 374 → 400
× 95 → 100
40,000

9. 33 → 30
× 25 → 30
900

10. 724 → 700
× 41 → 40
28,000

11. 5,156 → 5000
× 613 → 600
3,000,000

12. 979 → 1000
× 253 → 300
300,000

13. 96 → 100
× 32 → 30
192 3,000
2880
3,072

14. 21 → 20
× 47 → 50
147 1,000
840
987

15. 109 → 100
× 26 → 30
654 3,000
2180
2,834

16. 862 → 900
× 51 → 50
862 45,000
43100
43,962

17. 13 → 10
× 24 → 20
200
estimated ounces

18. 18 → 20
× 52 → 50
1,000
estimated flyers

Lesson 13 ······▸ MULTIPLICATION REVIEW (PAGES 26–28)

1. 24	**2.** 27	**3.** 2	**4.** 0	**5.** 376	**6.** 78
7. 299	**8.** 1,008	**9.** 1,786	**10.** 1,920	**11.** 16,192	**12.** 19,162
13. 6,660	**14.** 174,000	**15.** 575,100	**16.** 858,568	**17.** 768	**18.** 8,127
19. 153,966	**20.** 24,534	**21.** 6,000	**22.** 378,840	**23.** 4,800	**24.** 60,750

25. 62 → 60
× 47 → 50
2,914 3,000

26. 92 → 90
× 38 → 40
3,496 3,600

27. 56 → 60
× 19 → 20
1,064 1,200

28. 107 → 100
× 83 → 80
8,881 8,000

29. $663 \rightarrow 700$; $\times 521 \rightarrow 500$; $345{,}423$ $350{,}000$

30. $491 \rightarrow 500$; $\times 76 \rightarrow 80$; $37{,}316$ $40{,}000$

31. $64 \rightarrow 60$; $\times 15 \rightarrow 20$; 960 $1{,}200$

32. $514 \rightarrow 500$; $\times 439 \rightarrow 400$; $225{,}646$ $200{,}000$

33. $287 \rightarrow 300$; $\times 64 \rightarrow 60$; $18{,}368$ $18{,}000$

34. 6 cups

35. Both 16 and 18 round to 20. $20 \times 30 = 600$ estimated hours.

36. 28×20 = 560 miles Yes, he could make the trip on one tank.

37. $129

38. $2,844

39. 45,000 meters

40. $365 × 61 = $22,265

Lesson 14 ······▶ PROBLEM SOLVING WITH ADDITION, SUBTRACTION, AND MULTIPLICATION (PAGES 30–31)

1.

Question	Information	Method	Process	Answer
How much was left to pay?	cost of new refrigerator: $719 trade in: $75	Subtract	$719 − 75 $644	They had $644 left to pay.

2.

Question	Information	Method	Process	Answer
How many pictures did she take?	3 rolls with 24 2 rolls with 36	Multiply Add	24 × 3 = 72 36 × 2 = 72 72 + 72 = 144	She took 144 pictures.

3.

Question	Information	Method	Process	Answer
What will be cost of glasses after rebate?	$15 rebate $89 for glasses	Subtract	$89 −15 $74	The cost of the glasses will be $74.

4.

Question	Information	Method	Process	Answer
Estimate how many people go through line?	20 check out lines 18 customers per hour	Round Multiply	$20 \rightarrow 20$ $\times 18 \rightarrow 20$ 400	Approximately 400 customers go through line.

5.

Question	Information	Method	Process	Answer
About how many customers go through the express line line in 7 hours?	31 customers per hour 7 hours	Multiply	31 $\times 7$ 217	217 customers go through the express line in 7 hours.

6. $24

7. The first job pays $240 per week. The second job pays $245 per week. The second job pays more money.

8. $16 × 2 = $32 The pants will cost $32. $45 − $32 = $13 She will have $13 left.

Lesson 15 ······➤ WHAT IS DIVISION? (PAGES 32–33)

1. $(9 - 3 = 6)$ $(6 - 3 = 3)$ $(3 - 3 = 0)$ $9 \div 3 = 3$ $\begin{array}{r}3\\3\overline{)9}\end{array}$

2. $(10 - 5 = 5)$ $(5 - 5 = 0)$ $10 \div 5 = 2$ $\begin{array}{r}2\\5\overline{)10}\end{array}$

3. $(8 - 2 = 6)$ $(6 - 2 = 4)$ $(4 - 2 = 2)$ $(2 - 2 = 0)$ $8 \div 2 = 4$ $\begin{array}{r}4\\2\overline{)8}\end{array}$

4. $(15 - 5 = 10)$ $(10 - 5 = 5)$ $(5 - 5 = 0)$ $15 \div 5 = 3$ $\begin{array}{r}3\\5\overline{)15}\end{array}$

5. $(24 - 8 = 16)$ $(16 - 8 = 8)$ $(8 - 8 = 0)$ $24 \div 8 = 3$ $\begin{array}{r}3\\8\overline{)24}\end{array}$

6. $35 \div 7 = 5$ 7. $14 \div 2 = 7$ 8. $27 \div 3 = 9$ 9. $9 \div 9 = 1$ 10. $21 \div 7 = 3$

11. $36 \div 6 = 6$ 12. $\begin{array}{r}5\\5\overline{)25}\end{array}$ 13. $\begin{array}{r}4\\9\overline{)36}\end{array}$ 14. $\begin{array}{r}7\\4\overline{)28}\end{array}$ 15. $\begin{array}{r}2\\6\overline{)12}\end{array}$

16. $\begin{array}{r}6\\4\overline{)24}\end{array}$ 17. $\begin{array}{r}1\\9\overline{)9}\end{array}$ 18. $\begin{array}{r}6\\7\overline{)42}\end{array}$ 19. $\begin{array}{r}4\\2\overline{)8}\end{array}$ players

Lesson 16 ······➤ DIVISION FACTS (PAGES 34–35)

1. $\begin{array}{r}7\\1\overline{)7}\end{array}$ $\begin{array}{r}3\\1\overline{)3}\end{array}$ $\begin{array}{r}8\\1\overline{)8}\end{array}$ $\begin{array}{r}9\\1\overline{)9}\end{array}$ $\begin{array}{r}1\\1\overline{)1}\end{array}$ $\begin{array}{r}4\\1\overline{)4}\end{array}$ $\begin{array}{r}5\\1\overline{)5}\end{array}$ $\begin{array}{r}6\\1\overline{)6}\end{array}$ $\begin{array}{r}0\\1\overline{)0}\end{array}$ $\begin{array}{r}2\\1\overline{)2}\end{array}$

2. $\begin{array}{r}4\\2\overline{)8}\end{array}$ $\begin{array}{r}2\\2\overline{)4}\end{array}$ $\begin{array}{r}7\\2\overline{)14}\end{array}$ $\begin{array}{r}3\\2\overline{)6}\end{array}$ $\begin{array}{r}5\\2\overline{)10}\end{array}$ $\begin{array}{r}8\\2\overline{)16}\end{array}$ $\begin{array}{r}0\\2\overline{)0}\end{array}$ $\begin{array}{r}6\\2\overline{)12}\end{array}$ $\begin{array}{r}9\\2\overline{)18}\end{array}$ $\begin{array}{r}1\\2\overline{)2}\end{array}$

3. $\begin{array}{r}3\\3\overline{)9}\end{array}$ $\begin{array}{r}2\\3\overline{)6}\end{array}$ $\begin{array}{r}5\\3\overline{)15}\end{array}$ $\begin{array}{r}8\\3\overline{)24}\end{array}$ $\begin{array}{r}4\\3\overline{)12}\end{array}$ $\begin{array}{r}9\\3\overline{)27}\end{array}$ $\begin{array}{r}7\\3\overline{)21}\end{array}$ $\begin{array}{r}1\\3\overline{)3}\end{array}$ $\begin{array}{r}6\\3\overline{)18}\end{array}$ $\begin{array}{r}4\\3\overline{)12}\end{array}$

4. $\begin{array}{r}9\\4\overline{)36}\end{array}$ $\begin{array}{r}0\\4\overline{)0}\end{array}$ $\begin{array}{r}5\\4\overline{)20}\end{array}$ $\begin{array}{r}7\\4\overline{)28}\end{array}$ $\begin{array}{r}3\\4\overline{)12}\end{array}$ $\begin{array}{r}8\\4\overline{)32}\end{array}$ $\begin{array}{r}4\\4\overline{)16}\end{array}$ $\begin{array}{r}2\\4\overline{)8}\end{array}$ $\begin{array}{r}6\\4\overline{)24}\end{array}$ $\begin{array}{r}1\\4\overline{)4}\end{array}$

5. $\begin{array}{r}1\\5\overline{)5}\end{array}$ $\begin{array}{r}4\\5\overline{)20}\end{array}$ $\begin{array}{r}7\\5\overline{)35}\end{array}$ $\begin{array}{r}0\\5\overline{)0}\end{array}$ $\begin{array}{r}9\\5\overline{)45}\end{array}$ $\begin{array}{r}3\\5\overline{)15}\end{array}$ $\begin{array}{r}8\\5\overline{)40}\end{array}$ $\begin{array}{r}6\\5\overline{)30}\end{array}$ $\begin{array}{r}5\\5\overline{)25}\end{array}$ $\begin{array}{r}2\\5\overline{)10}\end{array}$

6. $\begin{array}{r}3\\6\overline{)18}\end{array}$ $\begin{array}{r}6\\6\overline{)36}\end{array}$ $\begin{array}{r}2\\6\overline{)12}\end{array}$ $\begin{array}{r}7\\6\overline{)42}\end{array}$ $\begin{array}{r}1\\6\overline{)6}\end{array}$ $\begin{array}{r}4\\6\overline{)24}\end{array}$ $\begin{array}{r}8\\6\overline{)48}\end{array}$ $\begin{array}{r}9\\6\overline{)54}\end{array}$ $\begin{array}{r}0\\6\overline{)0}\end{array}$ $\begin{array}{r}5\\6\overline{)30}\end{array}$

7. $\begin{array}{r}0\\7\overline{)0}\end{array}$ $\begin{array}{r}5\\7\overline{)35}\end{array}$ $\begin{array}{r}4\\7\overline{)28}\end{array}$ $\begin{array}{r}1\\7\overline{)7}\end{array}$ $\begin{array}{r}6\\7\overline{)42}\end{array}$ $\begin{array}{r}3\\7\overline{)21}\end{array}$ $\begin{array}{r}7\\7\overline{)49}\end{array}$ $\begin{array}{r}2\\7\overline{)14}\end{array}$ $\begin{array}{r}8\\7\overline{)56}\end{array}$ $\begin{array}{r}9\\7\overline{)63}\end{array}$

8. $\begin{array}{r}9\\8\overline{)72}\end{array}$ $\begin{array}{r}2\\8\overline{)16}\end{array}$ $\begin{array}{r}0\\8\overline{)0}\end{array}$ $\begin{array}{r}4\\8\overline{)32}\end{array}$ $\begin{array}{r}1\\8\overline{)8}\end{array}$ $\begin{array}{r}6\\8\overline{)48}\end{array}$ $\begin{array}{r}7\\8\overline{)56}\end{array}$ $\begin{array}{r}8\\8\overline{)64}\end{array}$ $\begin{array}{r}5\\8\overline{)40}\end{array}$ $\begin{array}{r}3\\8\overline{)24}\end{array}$

9. $\begin{array}{r}3\\9\overline{)27}\end{array}$ $\begin{array}{r}4\\9\overline{)36}\end{array}$ $\begin{array}{r}6\\9\overline{)54}\end{array}$ $\begin{array}{r}2\\9\overline{)18}\end{array}$ $\begin{array}{r}9\\9\overline{)81}\end{array}$ $\begin{array}{r}0\\9\overline{)0}\end{array}$ $\begin{array}{r}8\\9\overline{)72}\end{array}$ $\begin{array}{r}1\\9\overline{)9}\end{array}$ $\begin{array}{r}7\\9\overline{)63}\end{array}$ $\begin{array}{r}5\\9\overline{)45}\end{array}$

10. $3 \times 2 = 6$ $2 \times 3 = 6$ $6 \div 3 = 2$ $6 \div 2 = 3$

11. $7 \times 8 = 56$ $8 \times 7 = 56$ $56 \div 7 = 8$ $56 \div 8 = 7$

12. $4 \times 9 = 36$ $9 \times 4 = 36$ $36 \div 4 = 9$ $36 \div 9 = 4$

13. $5 \times 1 = 5$ $1 \times 5 = 5$ $5 \div 5 = 1$ $5 \div 1 = 5$

14. $6 \times 4 = 24$ $4 \times 6 = 24$ $24 \div 6 = 4$ $24 \div 4 = 6$

15. $7 \times 5 = 35$ $5 \times 7 = 35$ $35 \div 7 = 5$ $35 \div 5 = 7$

16. $8 \times 3 = 24$ $3 \times 8 = 24$ $24 \div 8 = 3$ $24 \div 3 = 8$

17. $2 \times 6 = 12$ $6 \times 2 = 12$ $12 \div 2 = 6$ $12 \div 6 = 2$

18. $9 \times 6 = 54$ $6 \times 9 = 54$ $54 \div 9 = 6$ $54 \div 6 = 9$

19. $32 \div 8 = 4$ $32 \div 4 = 8$
20. $45 \div 5 = 9$ $45 \div 9 = 5$
21. $6 \div 2 = 3$ $6 \div 3 = 2$

22. $18 \div 6 = 3$ $18 \div 3 = 6$
23. $28 \div 7 = 4$ $28 \div 4 = 7$
24. $56 \div 8 = 7$ $56 \div 7 = 8$

25. $36 \div 9 = 4$ $36 \div 4 = 9$
26. $30 \div 6 = 5$ $30 \div 5 = 6$
27. $14 \div 7 = 2$ $14 \div 2 = 7$

28. $32 \div \underline{4} = 8$
29. $56 \div \underline{8} = 7$
30. $40 \div \underline{5} = 8$

31. $81 \div \underline{9} = 9$
32. $6 \div \underline{2} = 3$
33. $54 \div \underline{9} = 6$

34. $24 \div \underline{3} = 8$
35. $42 \div \underline{7} = 6$
36. $72 \div \underline{9} = 8$

37. $64 \div \underline{8} = 8$
38. $18 \div \underline{6} = 3$
39. $9 \div \underline{9} = 1$

40. $24 \div \underline{8} = 3$
41. $27 \div \underline{3} = 9$
42. $25 \div \underline{5} = 5$

Lesson 17 ······➤ CHECKING DIVISION BY MULTIPLICATION (PAGES 37)

1. $7 \times 6 = 42$
$42 \div 6 = 7$

2. $6 \times 5 = 30$
$30 \div 5 = 6$

check | *check*

3. $16 \div 8 = 2$ $2 \times 8 = 16$

4. $21 \div 3 = 7$ $3 \times 7 = 21$

5. $7 \times 7 = 49$ $49 \div 7 = 7$

6. $9 \times 4 = 36$ $36 \div 4 = 9$

7. $4 \times 3 = 12$ $12 \div 3 = 4$

8. $6 \times 5 = 30$ $30 \div 5 = 6$

9. $2 \times 5 = 10$ $10 \div 5 = 2$

10. $5 \times 8 = 40$ $40 \div 8 = 5$

check | *check*

11. $25 \div 5 = 5$ $5 \times 5 = 25$

12. $10 \div 2 = 5$ $5 \times 2 = 10$

check | *check*

13. $64 \div 8 = 8$ $8 \times 8 = 64$

14. $8 \div 1 = 8$ $8 \times 1 = 8$

check | *check*

15. $28 \div 4 = 7$ $7 \times 4 = 28$

16. $30 \div 6 = 5$ $5 \times 6 = 30$

check | *check*

17. $18 \div 3 = 6$ $6 \times 3 = 18$

18. $54 \div 9 = 6$ $6 \times 9 = 54$

19. $21 \div 7 = 3$ $3 \times 7 = 21$

20. $30 \div 6 = 5 $6 \times $5 = 30

Lesson 18 ······➤ ONE-DIGIT DIVISION (PAGE 39)

1.
24
4)96
8
16
16
0

2.
173
5)865
5
36
35
15
15
0

3.
21
3)63
6
03
3
0

4.
135
7)945
7
24
21
35
35
0

5.
28
3)84
6
24
24
0

6.
12
6)72
6
12
12
0

7.
136
7)952
7
25
21
42
42
0

8.
14
6)84
6
24
24
0

9. 12
8)96
8
16
16
0

10. 311
3)933
9
03
3
03
3
0

11. 22
4)88
8
08
8
0

12. 16
5)80
5
30
30
0

13. 179
4)716
4
31
28
36
36
0

14. 342
2)684
6
08
8
04
4
0

15. 18
3)54
3
24
24
0

16. 13
5)65
5
15
15
0

17. 115
5)575
5
07
5
25
25
0

18. 12
7)84
7
14
14
0

19. 470
2)940
8
14
14
0

20. 22
3)66
6
06
6
0

21. 157
4)628
4
22
20
28
28
0

22. 11
9)99
9
09
9
0

23.

MARY JEAN
122
3)366
3
06
6
06
6
0

MEL
125
3)375
3
07
6
15
15
0

KAREN
127
3)381
3
08
6
21
21
0

LARRY
114
3)342
3
04
3
12
12
0

Lesson 19 ······► ONE-DIGIT DIVISION WITH REMAINDERS (PAGE 41)

1. 14 R3
5)73

2. 18 R5
7)131

3. 8 R1
6)49

4. 129 R1
4)517

5. 45 R3
5)228

6. 11 R1
8)89

7. 243 R1
2)487

8. 57 R3
8)459

9. 71 R1
7)498

10. 13 R5
6)83

11. 97 R2
5)487

12. 21 R1
7)148

13. 45 R2
7)317

14. 82 R2
5)412

15. 281 R1
2)563

16. 4 R5
6)29

17. 115 R1
8)921

18. 70 R1
3)211

19. 275 R1 miles
3)826

20. $ 57 R1
2)$115

21. $21 R3
4)$87

Lesson 20 ······▸ MORE ABOUT DIVIDING AND CHECKING (PAGE 43)

1.
19 R2
4)78
4
38
36
2

check
19
× 4
76
+ 2
78

2.
98 R1
5)491
45
41
40
1

check
98
× 5
490
+ 1
491

3.
146 R1
4)585
4
18
16
25
24
1

check
146
× 4
584
+ 1
585

4.
29 R4
8)236
16
76
72
4

check
29
× 8
232
+ 4
236

5.
7 R3
7)52
49
3

check
7
× 7
49
+ 3
52

6.
25 R1
3)76
6
16
15
1

check
25
× 3
75
+ 1
76

7.
82 R2
5)412
40
12
10
2

check
82
× 5
410
+ 2
412

8.
116 R3
4)467
4
06
4
27
24
3

check
116
× 4
464
+ 3
467

9.
25 R4
6)154
12
34
30
4

check
25
× 6
150
+ 4
154

10.
42 R3
9)381
36
21
18
3

check
42
× 9
378
+ 3
381

11.
912 R1
3)2,737
27
03
3
07
6
1

check
912
× 3
2,736
+ 1
2,737

12.
1097 R2
6)6,584
6
05
0
58
54
44
42
2

check
1,097
× 6
6,582
+ 2
6,584

13. 476 R2
9)4,286
36
68
63
56
54

check
476
× 9
4,284
+ 2
4,286

14. 855 R1 miles
4)3,421
32
22
20
21
20
1

check
855
× 4
3,420
+ 1
3,421

15. 29 R5 miles
8)237
16
77
72
5

check
29
× 8
232
+ 5
237

16. 380 R2 Sets
4)1,522
12
32
32
02
0
2

check
380
× 4
1,520
+ 2
1,522

Lesson 21 ······► TWO-DIGIT DIVISION (PAGES 45–46)

1. 36)72 = 2	**2.** 53)583 = 11	**3.** 26)338 = 13	**4.** 31)837 = 27	**5.** 13)1,586 = 122
6. 24)912 = 38	**7.** 28)896 = 32	**8.** 49)882 = 18	**9.** 22)1,958 = 89	**10.** 45)1,665 = 37
11. 63)945 = 15	**12.** 16)384 = 24	**13.** 33)792 = 24	**14.** 45)1,575 = 35	**15.** 82)2,952 = 36
16. 64)3,264 = 51	**17.** 39)1,053 = 27	**18.** 48)384 = 8	**19.** 51)459 = 9	**20.** 67)5,427 = 81

21. 5 display cases **22.** 28 display columns **23.** 34 copies

Lesson 22 ······► TWO-DIGIT DIVISION WITH REMAINDERS (PAGES 48–49)

1. 32)78 = 2 R14	**2.** 41)359 = 8 R31	**3.** 60)912 = 15 R12	**4.** 75)461 = 6 R11	**5.** 47)83 = 1 R36
6. 33)563 = 17 R2	**7.** 52)189 = 3 R33	**8.** 16)742 = 46 R6	**9.** 85)104 = 1 R19	**10.** 30)51 = 1 R21
11. 62)332 = 5 R22	**12.** 41)769 = 18 R31	**13.** 29)506 = 17 R13	**14.** 55)391 = 7 R6	**15.** 46)196 = 4 R12
16. 82)287 = 3 R41	**17.** 13)74 = 5 R9	**18.** 24)482 = 20 R2	**19.** 68)923 = 13 R39	**20.** 57)644 = 11 R17

21. 16 banners. There will be 22 feet of plastic left over.

22. 13 groups of tickets with 1 ticket remaining.

23. 10 packages. The vendor needs 9 full packages and part of a tenth package.

Lesson 23 ······► DIVISION WITH ZERO (PAGE 51)

1. $\begin{array}{r}66\\5\overline{)330}\end{array}$	**2.** $\begin{array}{r}151\text{ R3}\\4\overline{)607}\end{array}$	**3.** $\begin{array}{r}571\text{ R3}\\7\overline{)4{,}000}\end{array}$	**4.** $\begin{array}{r}1{,}686\text{ R2}\\3\overline{)5{,}060}\end{array}$
5. $\begin{array}{r}167\text{ R2}\\3\overline{)503}\end{array}$	**6.** $\begin{array}{r}1{,}334\text{ R4}\\6\overline{)8{,}008}\end{array}$	**7.** $\begin{array}{r}61\text{ R3}\\5\overline{)308}\end{array}$	**8.** $\begin{array}{r}71\text{ R1}\\9\overline{)640}\end{array}$
9. $\begin{array}{r}1{,}533\text{ R2}\\6\overline{)9{,}200}\end{array}$	**10.** $\begin{array}{r}57\text{ R1}\\7\overline{)400}\end{array}$	**11.** $\begin{array}{r}150\text{ R6}\\8\overline{)1{,}206}\end{array}$	**12.** $\begin{array}{r}2{,}697\\3\overline{)8{,}091}\end{array}$
13. $\begin{array}{r}117\\6\overline{)702}\end{array}$	**14.** $\begin{array}{r}411\text{ R3}\\5\overline{)2{,}058}\end{array}$	**15.** $\begin{array}{r}1{,}001\\3\overline{)3{,}003}\end{array}$	**16.** $\begin{array}{r}7\text{ R4}\\8\overline{)60}\end{array}$
17. 16 table covers	**18.** 10 hours	**19.** $14	

Lesson 24 ······► DIVISION BY NUMBERS ENDING IN ZERO (PAGES 52–53)

1. $\begin{array}{r}4\\10\overline{)40}\end{array}$	**2.** $\begin{array}{r}10\\60\overline{)600}\end{array}$	**3.** $\begin{array}{r}20\\40\overline{)800}\end{array}$	**4.** $\begin{array}{r}33\\300\overline{)9{,}900}\end{array}$
5. $\begin{array}{r}6\\50\overline{)300}\end{array}$	**6.** $\begin{array}{r}70\\30\overline{)2{,}100}\end{array}$	**7.** $\begin{array}{r}58\\10\overline{)580}\end{array}$	**8.** $\begin{array}{r}7\\70\overline{)490}\end{array}$
9. $\begin{array}{r}41\\20\overline{)820}\end{array}$	**10.** $\begin{array}{r}2\\80\overline{)160}\end{array}$	**11.** $\begin{array}{r}22\\200\overline{)4{,}400}\end{array}$	**12.** $\begin{array}{r}60\\100\overline{)6{,}000}\end{array}$
13. $\begin{array}{r}10\text{ R6}\\20\overline{)206}\end{array}$	**14.** $\begin{array}{r}14\text{ R9}\\30\overline{)429}\end{array}$	**15.** $\begin{array}{r}8\text{ R37}\\60\overline{)517}\end{array}$	**16.** $\begin{array}{r}70\text{ R8}\\10\overline{)708}\end{array}$
17. $\begin{array}{r}6\text{ R31}\\50\overline{)331}\end{array}$	**18.** $\begin{array}{r}2\text{ R42}\\400\overline{)842}\end{array}$	**19.** $\begin{array}{r}22\text{ R11}\\300\overline{)6{,}611}\end{array}$	**20.** $\begin{array}{r}136\text{ R3}\\70\overline{)9{,}523}\end{array}$
21. 46 packages	**22.** 10 bundles	**23.** 30 hours	

Lesson 25 ······► ROUNDING AND ESTIMATING IN DIVISION (PAGES 54–55)

1. $37\overline{)4{,}102} \longrightarrow$ *round* $\begin{array}{r}100\\40\overline{)4000}\end{array}$	**2.** $142\overline{)286} \longrightarrow \begin{array}{r}3\\100\overline{)300}\end{array}$
3. $\begin{array}{r}34\text{ R4}\\22\overline{)752}\end{array} \longrightarrow \begin{array}{r}40\\20\overline{)800}\end{array}$	**4.** $\begin{array}{r}59\text{ R17}\\68\overline{)4{,}029}\end{array} \longrightarrow \begin{array}{r}57\\70\overline{)4000}\end{array}$
5. $53\overline{)224} \longrightarrow \begin{array}{r}40\\50\overline{)200}\end{array}$	**6.** $49\overline{)841} \longrightarrow \begin{array}{r}16\\50\overline{)800}\end{array}$
7. $12\overline{)163} \longrightarrow \begin{array}{r}20\\10\overline{)200}\end{array}$	**8.** $55\overline{)977} \longrightarrow \begin{array}{r}16\\60\overline{)1{,}000}\end{array}$

9. $16\overline{)761} \longrightarrow \begin{array}{r} 40 \\ 20\overline{)800} \end{array}$

10. $91\overline{)1{,}845} \longrightarrow \begin{array}{r} 22 \\ 90\overline{)2{,}000} \end{array}$

11. $112\overline{)1{,}390} \longrightarrow \begin{array}{r} 10 \\ 100\overline{)1000} \end{array}$

12. $247\overline{)8{,}006} \longrightarrow \begin{array}{r} 40 \\ 200\overline{)8{,}000} \end{array}$

13. $486\overline{)1{,}660} \longrightarrow \begin{array}{r} 4 \\ 500\overline{)2000} \end{array}$

14. $\begin{array}{r} 41 \text{ R1} \\ 12\overline{)493} \end{array} \longrightarrow \begin{array}{r} 50 \\ 10\overline{)500} \end{array}$

15. $\begin{array}{r} 30 \text{ R10} \\ 25\overline{)760} \end{array} \longrightarrow \begin{array}{r} 26 \\ 30\overline{)800} \end{array}$

16. $34\overline{)590} \longrightarrow \begin{array}{r} 20 \\ 30\overline{)600} \end{array}$ estimated minutes

17. $720\overline{)8{,}640} \longrightarrow \begin{array}{r} 12 \\ 700\overline{)9{,}000} \end{array}$ estimated cans

Lesson 26 → DIVISION REVIEW (PAGES 56–57)

1. $\begin{array}{r} 5 \\ 7\overline{)35} \end{array}$

2. $\begin{array}{r} 4 \\ 4\overline{)16} \end{array}$

3. $\begin{array}{r} 23 \\ 6\overline{)138} \end{array}$

4. $\begin{array}{r} 157 \\ 3\overline{)471} \end{array}$

5. $\begin{array}{r} 9 \\ 5\overline{)45} \end{array}$

6. $\begin{array}{r} 8 \\ 9\overline{)72} \end{array}$

7. $\begin{array}{r} 8 \\ 8\overline{)64} \end{array}$

8. $\begin{array}{r} 13 \\ 2\overline{)26} \end{array}$

9. $\begin{array}{r} 74 \\ 2\overline{)148} \end{array}$

10. $\begin{array}{r} 37 \\ 9\overline{)333} \end{array}$

11. $\begin{array}{r} 123 \\ 8\overline{)984} \end{array}$

12. $\begin{array}{r} 52 \\ 5\overline{)260} \end{array}$

13. $\begin{array}{r} 33 \\ 27\overline{)891} \end{array}$

14. $\begin{array}{r} 349 \text{ R3} \\ 5\overline{)1{,}748} \end{array}$

15. $\begin{array}{r} 537 \text{ R3} \\ 7\overline{)3{,}762} \end{array}$

16. $\begin{array}{r} 2 \text{ R5} \\ 55\overline{)115} \end{array}$

17. $\begin{array}{r} 60 \\ 5\overline{)300} \end{array}$

18. $\begin{array}{r} 500 \text{ R6} \\ 8\overline{)4{,}006} \end{array}$

19. $\begin{array}{r} 175 \\ 6\overline{)1{,}050} \end{array}$

20. $\begin{array}{r} 4 \text{ R4} \\ 75\overline{)304} \end{array}$

21. $\begin{array}{r} 21 \text{ R2} \\ 70\overline{)1{,}472} \end{array}$

22. $\begin{array}{r} 98 \text{ R55} \\ 60\overline{)5{,}935} \end{array}$

23. $\begin{array}{r} 4 \\ 400\overline{)1{,}600} \end{array}$

24. $\begin{array}{r} 206 \text{ R24} \\ 30\overline{)6{,}204} \end{array}$

25. $\begin{array}{r} 87 \\ 6\overline{)522} \end{array}$ *check* $\begin{array}{r} 87 \\ \underline{\times\ 6} \\ 522 \end{array}$

26. $\begin{array}{r} 93 \text{ R4} \\ 7\overline{)655} \end{array}$ $\begin{array}{r} 93 \\ \underline{\times\ 7} \\ 651 \\ \underline{+\ 4} \\ 655 \end{array}$

27. $\begin{array}{r} 12 \text{ R16} \\ 83\overline{)1{,}012} \end{array}$ $\begin{array}{r} 83 \\ \underline{\times\ 12} \\ 166 \\ \underline{830} \\ 996 \\ \underline{+\ 16} \\ 1{,}012 \end{array}$

28. $\begin{array}{r} 8 \text{ R28} \\ 54\overline{)460} \end{array}$ $\begin{array}{r} 54 \\ \underline{\times\ 8} \\ 432 \\ \underline{+\ 28} \\ 460 \end{array}$

29. $\begin{array}{r} 5 \text{ R27} \\ 31\overline{)182} \end{array} \longrightarrow \begin{array}{r} 6 \\ 30\overline{)200} \end{array}$

30. $\begin{array}{r} 10 \text{ R33} \\ 49\overline{)523} \end{array} \longrightarrow \begin{array}{r} 10 \\ 50\overline{)500} \end{array}$

31. $\begin{array}{r} 31 \text{ R180} \\ 300\overline{)9{,}480} \end{array} \longrightarrow \begin{array}{r} 30 \\ 300\overline{)9{,}000} \end{array}$

32. $\begin{array}{r} 9 \text{ R24} \\ 518\overline{)4{,}686} \end{array} \longrightarrow \begin{array}{r} 10 \\ 500\overline{)5{,}000} \end{array}$

33. $\begin{array}{r} 42 \text{ R1} \\ 23\overline{)967} \end{array} \longrightarrow \begin{array}{r} 50 \\ 20\overline{)1{,}000} \end{array}$

34. $\begin{array}{r} 39 \text{ R22} \\ 64\overline{)2{,}518} \end{array} \longrightarrow \begin{array}{r} 50 \\ 60\overline{)3{,}000} \end{array}$

35. 23 shelves Felipe will fill 22 shelves completely and part of the next shelf.

36. 563 seats **37.** About 45 parts **38.** 19 weeks

39. \$380 ÷ 2 *roommates* = \$190 \$380 ÷ 3 roommates = \$126 R\$2 **40.** 14 people per team

Lesson 27 ······➤ MULTIPLICATION AND DIVISION REVIEW (PAGE 58)

1. 296 **2.** 70 **3.** 34 **4.** 665 **5.** 768 **6.** 153

7. 39,483 **8.** 128 R 2 **9.** 52 **10.** 49,840 **11.** 44 **12.** 22 R 17

13. 31,746 **14.** 31 R 8 **15.** 8 R 31 **16.** 35,358 **17.** 4,767,455 **18.** 2,066 R 1

19. 83,200 **20.** 2,463,500 **21.** 121 R 2 **22.** 3,600

23. 471 → 500; × 65 → × 70; 35,000

24. 57)6,324 → 60)6,000 = 100

25. 9)888 → 10)900 = 90

26. 932 → 900; × 456 → × 500; 450,000

27. 9,215 → 9,000; × 18 → × 20; 180,000

28. 210)2,340 → 200)2,000 = 10

	Exact Answer	Estimate		Exact Answer	Estimate
29.	115,661	120,000	**30.**	116 R 49	114
31.	431 R 18	500	**32.**	220,020	240,000
33.	748,956	800,000	**34.**	30,485	36,000

Lesson 28 ······➤ PROBLEM SOLVING WITH MULTIPLICATION AND DIVISION (PAGES 60–61)

1.

Question	Information	Method	Process	Answer
How far should they run in one month?	300 miles 3 months	Divide	100 3)300	They should run 100 miles a month.

2.

Question	Information	Method	Process	Answer
How long will it take her to run 3 miles?	8 minutes 3 miles	Multiply	8 ×3 24	It will take her 24 minutes.

3.

Question	Information	Method	Process	Answer
How many miles did she run?	ran 100 minutes 11 minutes a mile	Divide	9 R1 11)100	About 9 miles.

4.

Question	Information	Method	Process	Answer
What price did she pay per issue?	paid $24 12 issues	Divide	$2 12)$24	She paid $2 per issue.

5.

Question	Information	Method	Process	Answer
How many laps did she run for 3 miles.	14 laps equals 1 mile	Multiply	14 ×3 42	She ran 42 laps.

6. 375 meters Divide 1,500 meters by 4 minutes.

7. 13 sets of bleachers Divide 26 miles by 2 miles.

8. 9,300 meters Multiply 310 meters by 30 minutes.

Lesson 29 ······➤ PROBLEM-SOLVING REVIEW (PAGES 62–64)

1.

Question	Information	Method	Process	Answer
Estimate number who swam at pool.	12 days 5,800 people	Round Estimate (Divide)	600 12)5,800 → 10)6,000 -60	An estimated 600 people swam each day.

2.

Question	Information	Method	Process	Answer
Estimate number who swam at pool.	12 days 5,800 people	Round Estimate (Divide)	Step 1: 3 2)6 Step 2: $10 × 3 $30	An estimated 600 people swam each day.

3.

Question	Information	Method	Process	Answer
How much will the payments be?	ring price $436 down payment $100 12 monthly payments	Subtract Divide	Step 1: $436 − 100 $336 Step 2: 28 12)$336 24 96 96	monthly payments will be $28.

4.

Question	Information	Method	Process	Answer
What is cost for family of 3 children?	$90 for 1st child $75 for 2nd child $35 for additional children	Add	$90 $75 + $35 $200	Cost for 3 children is $200.

5.

Question	Information	Method	Process		Answer
How many baskets did Ellen have left over?	68 baskets 7 baskets on 8 shelves	Multiply Subtract	7 × 8 56	68 − 56 12	Ellen had 12 baskets left over.

6. $766   Add: $78 + $112 = $190   Subtract: $956 − $190 = $766
7. 100 feet   Multiply: 25 × 4 = 100
8. 71 hours   Add: 78 + 102 + 49 = 229   Subtract: 300 − 229 = 71
9. 11 months   Subtract: $525 − $200 = $325   Divide: $325 ÷ $30 = 10 R 25
10. $150   Multiply: $75 × 6 = $450   Subtract: $600 − $450 = $150

Lesson 30 ······► PERIMETER (PAGES 65–67)

1. 28 yards
30
+ 40
98 yards

2. 12 feet
12
40
+ 40
104 feet

3. 8 inches
8
12
+ 12
40 inches

4. 60 inches
60
80
+ 80
280 inches

5. 4 miles
2
4
+ 2
12 miles

6. 35 yards
15
+ 37
87 yards

7. 15 inches
30
15
17
+ 17
94 inches

8. 66 feet
66
58
+ 58
248 feet

9. 15 feet
10
15
+ 10
50 feet

10. 5 feet
7
+ 9
21 feet

11. 14 feet
14
12
+ 9
49 feet

12. 12,875 feet
9,500
7,130
900
+ 6,910
37,315 feet

13. 9 feet
4
+ 4
17 feet

Lesson 31 ······▸ AREA (PAGES 69–70)

1. 9 feet
× 6
54 square feet

2. 4 miles
× 4
16 square miles

3. 34 feet
× 17
238
340
578 square feet

4. 12 feet
× 9
108 square feet

5. 120 feet
× 80
9,600 square feet

6. 240 miles
× 240
9600
48000
57,600 square miles

7. 22 yards
× 37
154
660
814 square yards

8. 44 inches
× 11
44
440
484 square inches

9. 3 feet
× 7
21 square feet

10. 5 feet
× 5
25 square feet

11. 130 yards
× 70
9,100 square yards

12. 35 feet
× 50
1,750 square feet No, one bag will not be enough.

13. 25 feet
× 10
250 square feet

Lesson 32 ······▸ AVERAGES (PAGES 71–72)

1. 15 15 + 12 + 18 + 13 + 17 = 75 75 ÷ 5 = 15

2. $10 $10 + $8 + $12 + $10 = $40 $40 ÷ 4 = $10

3. 85 81 + 85 + 95 + 77 + 87 = 425 425 ÷ 5 = 85

4. 293 250 + 300 + 305 + 295 + 310 + 298 + =1,758 ÷ 6 = 293

5. 7 4 + 8 + 12 + 6 + 5 = 35 35 ÷ 5 = 7

6. 21,318 21,422 +21,224 + 21,422 + 21,204 = 85,272 85,272 ÷ 4 =21,318

7. $45 51 + 37 + 42 + 50 = $180 180 ÷ 4 = $45

8. 226 223 + 224 + 225 + 227 + 228 + 229 = 1,356 1,356 ÷ 6 = 226

9. 99° 98 + 99 + 103 + 100 + 96 + 96 =592 592 ÷ 6 = 98.6, rounds to 99

10. 13 40 ÷ 3 = 13.3, rounds to 13 pounds.

11. 36 years old 40 + 25 + 38 + 44 + 50 + 19 = 216 216 ÷ 6 = 36

12. $5 In tips $60 ÷ 12 = $5

13. 24 sec 23 + 24 + 25 = 72 72 ÷ 3 = 24

NOTES